THE RESILIENCE MACHINE

We live in a time where environmental pressures, social inequities and political derision are the backdrop of everyday life, and where resilience has become a routine prescription for coping with the conditions of modern existence. Drawing an analogy to Harvey Molotch's urban growth machine, this book explores different narratives of resilience and their policy and practice manifestations for cities, citizens and communities. It expands on the metaphor of the machine to show how resilience can be better understood as an assemblage.

Bringing together authors from multiple disciplines and different parts of the world, the book unmasks the often invisible effects of resilience strategies by examining ways in which neoliberal mentalities are fed through the rhetoric of resilience practices, policies and development projects. The contributing essays provide provocative accounts of several areas of inquiry, including biopolitics and smart bodies, resilient cities and communities, urban planning and disaster management, justice and vulnerability, and resistance to resilience. Holding out hope for critical potentials in 'resilience,' *The Resilience Machine* proposes to move beyond mechanisms of adaptation and into imagining what resilient life could look like in a more just, equitable and democratic world.

The Resilience Machine is a current, vital addition to resilience, community and urban scholarship.

Jim Bohland is Research Team Leader on the social and political dimensions of resilience at the Global Forum for Urban and Regional Resilience, and is Professor Emeritus in the School of Public and International Affairs at Virginia Tech. He is the former vice president and executive director of Virginia Tech's National Capital Region Operations and former director of the School of Public and International Affairs.

Simin Davoudi is Professor of Environmental Policy and Planning and Director of the Global Urban Research Unit at Newcastle University. She has held visiting professorships at universities in the USA, the Netherlands, Sweden, Australia and Finland. Her research centres on the politics of urban planning, securitisation of nature, resilience and governmentality of unknowns. Selected books include: *The Routledge Companion to Environmental Planning and Sustainability* (2019), *Justice and Fairness in the City* (2016), *Reconsidering Localism* (2015) and *Conceptions of Space and Place* (2009).

Jennifer Lawrence is a post-doctoral research associate with the Global Forum on Resilience, Virginia Tech. Her research explores the assemblage of extractive governance, by drawing out tensions between chronic and acute socio-environmental disasters. Her scholarship is conducted from a problem-centred, theory-driven methodology and highlights the intersection of economic systems, resource extraction and socio-environmental (in)justice. She is also the editor of *Biopolitical Disaster* (2017).

"A brilliant, empirically rich and theoretically inventive collection, which opens up new perspectives on resilience as a way of governing life itself. Organised around a novel conceptualisation of resilience as a 'machine,' the collection offers a politically incisive examination of the strategies, motivations, and logics that surround different enactments of resilience."

– Ben Anderson, Professor of Human Geography, Durham University, UK

"*The Resilience Machine* provides a unique, timely and indispensable critique of resilience narratives, policies and practices and how these have been shaped by dominant political and economic systems. While providing a much-needed critique of resilience practices which further promote neoliberal priorities, this book and its contributors also demonstrate how critical resilience thinking has the potential to produce desirable socio-spatial and environmental outcomes, providing a potential pathway for transformative and positive change. Bohland, Davoudi and Lawrence have assembled a volume that will have wide multidisciplinary appeal for students and researchers with interests in urban studies, disaster management, planning, community development and sustainability."

– Mark Scott, Professor of Planning, University College Dublin, Ireland

THE RESILIENCE MACHINE

Edited by Jim Bohland, Simin Davoudi and Jennifer Lawrence

NEW YORK AND LONDON

First published 2019
by Routledge
711 Third Avenue, New York, NY 10017

and by Routledge
2 Park Square, Milton Park, Abingdon, Oxon OX14 4RN

Routledge is an imprint of the Taylor & Francis Group, an informa business

© 2019 Taylor & Francis

The right of Jim Bohland, Simin Davoudi and Jennifer Lawrence to be identified as the authors of the editorial material, and of the authors for their individual chapters, has been asserted in accordance with sections 77 and 78 of the Copyright, Designs and Patents Act 1988.

All rights reserved. No part of this book may be reprinted or reproduced or utilised in any form or by any electronic, mechanical, or other means, now known or hereafter invented, including photocopying and recording, or in any information storage or retrieval system, without permission in writing from the publishers.

Trademark notice: Product or corporate names may be trademarks or registered trademarks, and are used only for identification and explanation without intent to infringe.

Library of Congress Cataloging in Publication Data
Names: Bohland, James R., editor. | Davoudi, Simin, editor. | Lawrence, Jennifer L., editor.
Title: The resilience machine / edited by Jim Bohland, Simin Davoudi and Jennifer Lawrence.
Description: New York : Routledge, 2019. | Includes index.
Identifiers: LCCN 2018024596| ISBN 9780815381129 (hardback) | ISBN 9780815381136 (pbk.)
Subjects: LCSH: City planning. | Sustainable development. | Emergency management. | Preparedness.
Classification: LCC HT166 .R428 2019 | DDC 307.1/216–dc23
LC record available at https://lccn.loc.gov/2018024596

ISBN: 978-0-8153-8112-9 (hbk)
ISBN: 978-0-8153-8113-6 (pbk)
ISBN: 978-1-351-21118-5 (ebk)

Typeset in Bembo
by Taylor & Francis Books

CONTENTS

List of figures   vii
List of contributors   viii
Acknowledgements   xii

Introduction   1
Jennifer Lawrence, Simin Davoudi and Jim Bohland

1 Anatomy of the Resilience Machine   12
   Simin Davoudi, Jennifer Lawrence and Jim Bohland

2 Securing the Imagination: The Politics of the Resilient Self   29
   Julian Reid

3 Designing 'Smart' Bodies: Molecular Manipulation as a
   Resilience-Building Strategy   43
   Rebecca J. Hester

4 Organising Community Resilience   62
   Chris Zebrowski and Daniel Sage

5 Rejecting and Recreating Resilience after Disaster   80
   Raven Cretney

6 The Resonance and Possibilities of *Community* Resilience   94
   Lauren Rickards, Martin Mulligan and Wendy Steele

7 Adaptation Machines, or the Biopolitics of Adaptation   110
   Kevin Grove and Jonathan Pugh

8 The Resilient City: Where Do We Go from Here? 125
   Peter Rogers

9 Towards a Critical Political Geography of Resilience Machines
   in Urban Planning 144
   Thilo Lang

10 Resilience and Justice: Planning for New York City 159
    Susan S. Fainstein

11 Seeking the Good (Enough) City 177
    Brendan Gleeson

12 Dismantling the Resilience Machine as a Restoration Engine 191
    Timothy W. Luke

Index *209*

FIGURES

6.1 The closed-open spectrum common to the diagrammatic form
of community and resilience 97
6.2 Speculative schematic of bounded and unbounded diagrams of
community intersecting with those of resilience 98
8.1 Urban systems model approach 132

CONTRIBUTORS

Jim Bohland is Research Team Leader on the social and political dimensions of resilience at the Global Forum for Urban and Regional Resilience, and is Professor Emeritus in the School of Public and International Affairs at Virginia Tech. He is the former vice president and executive director of Virginia Tech's National Capital Region Operations and former director of the School of Public and International Affairs.

Raven Cretney completed her doctorate at RMIT University in Melbourne, Australia. Her research approaches the politics of grassroots participation in environmental and social issues, particularly following disaster and crisis. Her previous work has explored radical articulations of resilience and the role of community-level organisation in disaster response and recovery.

Simin Davoudi is Professor of Environmental Policy and Planning and Director of the Global Urban Research Unit at Newcastle University. She has held visiting professorships at universities in the USA, the Netherlands, Sweden, Australia and Finland. Her research centres on the politics of urban planning, securitisation of nature, resilience and governmentality of unknowns. Selected books include: *The Routledge Companion to Environmental Planning and Sustainability* (2019), *Justice and Fairness in the City* (2016), *Reconsidering Localism* (2015) and *Conceptions of Space and Place* (2009).

Susan S. Fainstein is Senior Research Fellow and formerly professor of urban planning in the Harvard Design School. She previously taught at Columbia and Rutgers Universities and recently at the National University of Singapore. She received the Davidoff Book Award for *The Just City* from the Association of

Collegiate Schools of Planning. Her writings focus on planning theory, urban theory and urban redevelopment.

Brendan Gleeson is Professor of Urban Policy Studies and Director of the Melbourne Sustainable Society Institute at the University of Melbourne. His main research interests are urban political ecology, climate governance and urban social policy. His most recent book is *The Urban Condition* (2014).

Kevin Grove is Associate Professor in the Department of Global and Sociocultural Studies at Florida International University. His research examines the biopolitics of disaster resilience, with a focus on Caribbean emergency management and urban resilience planning in US cities. He is the author of *Resilience* and a number of articles in geography and interdisciplinary journals, including *Annals of the Association of American Geographers, Environment and Planning D: Society and Space, Geoforum, Antipode* and *Security Dialogue*.

Rebecca J. Hester is Assistant Professor in the Department of Science, Technology and Society at Virginia Polytechnic and State University. She received her PhD in Politics from the University of California Santa Cruz and was subsequently the Chancellor's Postdoctoral Fellow in Latino Studies at the University of Illinois Urbana-Champaign. Her research interests include biopolitics, biosecurity and biodefense, medical technologies, surveillance studies and posthumanism.

Thilo Lang is Head of Department at the Leibniz Institute for Regional Geography, Leipzig, Germany, and lecturer at Leipzig University. He completed his doctorate in human geography at the University of Potsdam, Germany and at Durham University, UK in 2008 and holds a diploma in spatial and environmental planning. His research interests include policy responses to socio-spatial polarisation, (social) innovation in remote areas as well as urban and regional change and resilience. He is co-editor of Palgrave's New Geographies of Europe book series.

Jennifer Lawrence is a post-doctoral research associate with the Global Forum on Resilience, Virginia Tech. Her research explores the assemblage of extractive governance, by drawing out tensions between chronic and acute socio-environmental disasters. Her scholarship is conducted from a problem-centred, theory-driven methodology and highlights the intersection of economic systems, resource extraction and socio-environmental (in)justice. She is also the editor of *Biopolitical Disaster* (2017).

Timothy W. Luke is University Distinguished Professor in the Department of Political Science at Virginia Polytechnic Institute and State University, Blacksburg, Virginia. His research focuses on critical theory, environmental politics, social and political theory in relation to global governance, political economy and cultural conflict, particularly with regard to the theories and practices of environmental management, resilience and sustainability.

Martin Mulligan is Associate Professor in Sustainability and Urban Planning and Centre for Urban Research in the School of Global, Urban and Social Studies at RMIT University, Melbourne. He has conducted research on disaster recovery, resilience and community 'wellbeing' in Australia, Sri Lanka and the UK and has co-authored papers on community resilience published in *International Planning Studies* and *Resilience: International Policies, Practices and Discourses*.

Jonathan Pugh, School of Geography, Politics and Sociology, Newcastle University, UK is an island studies scholar particularly associated with the 'relational turn' in island studies, and more recently island studies in the Anthropocene. He has more than 40 publications on island studies and related research interests in resilience, participatory development and radical politics, especially in the Caribbean. Jonathan has lectured on island studies at, among other universities, Taipei, Zurich, West Indies, London, Cornell, California, Princeton, Harvard and Virginia Tech.

Julian Reid is Chair and Professor of International Relations at the University of Lapland. His most recent book is *The Neoliberal Subject: Resilience, Adaptivity and Vulnerability* (co-authored with David Chandler, 2016). He is currently Principal Investigator of Indigeneity in Waiting, a research project funded by the Academy of Finland.

Lauren Rickards is Associate Professor at RMIT University, Melbourne where she co-leads the Climate Change and Resilience research programme of the Centre for Urban Research. She co-leads the Nature, Risk and Resilience study group of the Institute of Australian Geographers and is on the editorial board of *Resilience: International Policies, Practices and Discourses*. Her recent work includes analyses of ideas about resilience with Raven Cretney and Kevin Grove.

Peter Rogers is Senior Lecturer in Sociology at Macquarie University, Sydney. He has published widely on the theory and politics of space, in recent years focusing more specifically on genealogies of disaster resilience and the critique of neoliberal governance practices.

Daniel Sage is Senior Lecturer in Organisational Behaviour in the School of Business and Economics at Loughborough University, UK. His research concerns critical studies of organisations and management, especially related to power, workplace inequalities and organisational geographies. His contributions to these themes have developed in relation to various social challenges and contexts of organising, including critical infrastructure and community resilience.

Wendy Steele is Associate Professor in Cities, Sustainability and Planning co-located in the School of Global, Urban and Social Studies and the Centre for Urban Research at RMIT University, Melbourne. Her current research focuses on cities in a climate of change with an emphasis on critical urban governance. She is an international editorial board member for the journal *Urban Policy and Research*.

Chris Zebrowski is a lecturer in Politics and International Relations at Loughborough University, UK. His research has investigated the emergence of resilience discourses and their implications for the rationalities and practices of liberal emergency governance. He is an assistant editor of *Resilience: International Policies, Practices and Discourses* and author of *The Value of Resilience: Securing Life in the Twenty-First Century* (2016).

ACKNOWLEDGEMENTS

Myriad voices gave rise to this book. We would like to use this space to thank those who have been integral to the project by telling the story of how the book came about. At the centre of this story is the Global Forum on Urban and Regional Resilience and its predecessor the Center for Community Security and Resilience at Virginia Tech which played important coordinating and energising roles. In 2010, the Center began a series of workshops designed to explore new ideas, facilitate dialogue and debate some of the policy and ethical issues embedded in the then emerging concept called community resilience. Those conversations continued annually in Davos, Lugano, and Washington, DC. With the creation of the Forum, the scope of topics was broadened to include discussions of new perspectives on resilience – ethics, values, fear and political rationalities, to name a few. The diverse disciplinary and professional backgrounds of participants ensured that orthodox narratives on resilience were challenged, including most importantly the assumption that resilience was always a desirable community trait that should be advanced by all means possible. Rather, more critical views of resilience were raised and explanations posited for what seemed to be the mercurial spread of the resilience gospel. Of equal importance to the scholarly engagement, the workshops created a network of scholars with overlapping interests, some of whom resided in the Forum for short periods and helped shape the agenda for this book. We would like to extend our gratitude to everyone who participated in the debates and generously shared their ideas with us.

An important narrative in the initial critiques of resilience revolved around the question of whether resilience is a policy construct perfectly suited to a time of increased risk, or were agents using it to promote their agendas, and if the latter, who were they, and how did they function? From this evolved 'The Machine' as a frame for understanding the remarkable expansion of resilience intellectually and in public discourse. The rise of machine metaphor as a major theme within the Global Forum was a consequence of the interests of Paul Knox and Jim Bohland. From his scholarship on urbanisation, Knox viewed the adoption of resilience as a planning strategy analogous to what Molotch and Logan posited as the 'growth machine' that was responsible for shaping urban development. A machine in this context was conceptualised as a collection of actors pursuing a development agenda for their own financial gains. Bohland's interest in agency, values and policy provided an institutional lens to what was called the 'resilience machine'; an important research theme within the newly constituted Global Forum in 2013. Our special thanks goes to Paul for suggesting the title of the book and helping us with the development of its scope and outline and putting together a strong proposal to our publisher.

Two other voices essential to the origination of the resilience machine as a book were heard in 2014. In the autumn, the Global Forum hosted a workshop on the *Normative Dimensions of Resilience* in Washington, DC. Simin Davoudi and Jennifer Lawrence both presented work questioning the governmentality of resilience and the effects of resilience policies. In the following year, Simin came to the Forum as a visiting scholar and Jennifer joined as a post-doctoral research associate. While at the Forum, Simin's collaboration with Jim, Jennifer and Paul sharpened the arguments associated with the 'resilience machine' frame. An important decision in the book's evolution was the convening of a workshop on the 'resilience machine' in Washington, DC in the autumn of 2016. Here the contributors to the book gathered to share perspectives and provide important suggestions on themes and organisation of the book. As a consequence of that workshop, the writing of the book moved swiftly from concept to reality.

Special mention should be made of the role of the Global Forum under the leadership of the late Dr Charles Steger. Also, Paul Knox and Jim Bohland were instrumental in establishing the Forum, the visiting scholars' programme and the workshops that shaped the book. In addition to Simin Davoudi, Julian Reid and Jonathan Pugh were resident scholars in the Forum at different times during the book's progression. Tim Luke, Kevin Grove, Jennifer Lawrence and Simin Davoudi also presented the groundwork for the volume at the 2017 American Association of Geographers in Boston, MA. Finally, we would be remiss if we failed to mention Dr Faith Goodfellow at Newcastle University for her excellent editorial assistance and Krystal LaDuc at Routledge for her patience and helpful advice throughout the process.

INTRODUCTION

Jennifer Lawrence, Simin Davoudi and Jim Bohland

We are living in a time where environmental pressures, social inequities and political derision are the backdrop of everyday life. For many governmental institutions, organisations, cities, communities and individuals, resilience is a routine prescription for coping with the conditions of modern existence. The proliferation of resilience narratives, policies and practices spur the need for a clear understanding of ways in which global political and economic systems necessitate resilient subjectivities. Such considerations frame our exploration into narratives and imaginaries of resilience, inform questions about uneven geographies of accumulation by dispossession and shape our inquiry into the construction of what we call 'the resilience machine'.

In the chapter that follows, we draw an analogy to the urban growth machine, a powerful articulation of agents and mechanisms of economic growth expressed by Harvey Molotch (1976). Expanding on the metaphor of the machines, we explore how the resilience machine might be better understood as an 'assemblage'. Our contributors follow on exploring different ways in which neoliberal mentalities are fed through the rhetoric of resilience practices, policies and development projects and explore the often invisible effects of resilience strategies. While engaging with the metaphor of the machine, they have avoided being enticed into ascribing the deleterious effects of resilience to a nebulous system void of actors, agents and intentions. Rather, by elucidating the workings of resilience machines, they pick apart the elements of the assemblage. Predictably, we find that instrumentalist use of resilience is deployed and actively pursued by a spectrum of actors and institutions. We also find that the regressive effects of resilience policies have reverberating effects throughout communities. Nevertheless, we contend that space remains for resilience efforts to produce desirable outcomes, and that we should pay careful attention to the diverse values that drive resilience coalitions.

We reflect on the influence of resilience on a wide range of academic disciplines and interrogate the discursive entanglements of resilience in the material world. Confronting unreflective and alienating resilience initiatives, *The Resilience Machine* highlights the experiences of individuals, communities and cities in addressing the stressors of corporeal vulnerabilities, climate change, disaster risks and urbanisation. We believe that in these domains, and many others, there is much left to be understood about the functioning and experience of resilience as a contemporary subjectivity of life itself. The contributions to this volume animate the interconnectivity of resilience machines at multiple levels, and work to demonstrate their mutual constitution and reinforcement. Echoing a systems view of resilience put forth by Walker and Salt (2012), we take seriously the notion that resilience is often put forward as a strategy to govern life through its inherent complexity (Chandler, 2014). This perspective allows for systems and relationships that produce vulnerability and insecurity to be unveiled. Thus, entanglements of knowledge, power, agency and practice become central to understanding resilience machines and imagining alternative ways of being.

Examining the dependencies and relationships that enable unreflective modes of resilience policy means that we must confront underlying motivations. Here, *The Resilience Machine* challenges the dominant view of resilience as a desirable trait, while also recognising the importance of resilience as a potential means of transformative change. Spurred by cases of the lived experiences, the chapters in this volume question the systemic production of resilient subjectivity on multiple levels: the individual (both as a biological and social being), the community and the city. However, given the transmutable nature of resilience, the chapters do not easily lend themselves into such categorisation. Rather, the authors weave together elements from these levels to explore the multiplicity of its deployment in a number of interrelated, reinforcing and overlapping contexts. It is, therefore, more useful to consider these and other levels (e.g. global, ideological, molecular) as nodes through which resilience moves as a travelling concept. Interrogating the underlying logics of resilience is essential to understanding these scalar interdependencies and acknowledging that resilient assemblages can *both* promote *and* resist neoliberal rationalities.

As we note in Chapter 1, characterising assemblages as entities that display consistent structures with fuzzy boundaries provides a basis for understanding the transmutation of norms promoting specific resilience policies through, for example, a neoliberal frame into one that challenges and rejects prior norms and beliefs. Such cyborg functioning of resilience entails 'a legacy of systems theory, cybernetics thinking, and other ambivalent sources' (Åsberg, 2014). This fuzziness and proliferation necessitate relational thinking which has, indeed, informed the organisation of the chapters in this volume.

Organisation of the Anthology

The subjectivities, resilience and assemblages that come together are always in the process of becoming. This implies that resilient subjectivities, in their constant evolution, can 'comply or collude, subvert or resist discourses of resilience governing and disciplining arenas' (Aranda et al., 2012, page 554). Based on this perspective, we purposely resist segmenting the chapters into an artificial scalar categorisation. Instead, reflecting on the intersection of agents, discourses and effects as the strongest evidence of resilience machines, we have assembled the project in a way that allows the reader to explore the variegated mosaic of these vectors.

In this spirit, we begin the volume with 'Anatomy of the Resilience Machine'. The chapter is a call to action to make visible the interconnected imaginaries and technologies of resilience, and contextualise their effects within a material context. Offering a genealogy of the assemblages of resilience we draw together both the historical origins of the terms 'resilience' and 'machine' and question how the proliferation of resilience initiatives seamlessly blend into a wide range of disciplines, practices and policies. In this scene-setting chapter, we articulate how highly selected interpretations of resilience have become carriers of neoliberal strategies, which can lead to dangerous and deleterious effects and which have become subject to growing contestation. The chapter is a primer for contextualised explorations of resilience machines in the ensuing chapters.

Next, Julian Reid's chapter takes us on a theoretical journey exploring how the 'resilient self' is secured through imagination. Drawing connections to the circulation of state power, he indicates that collective imaginaries might be galvanised for the purpose of resilience building and thereby through manipulation. Reid asks us to consider how 'might we liberate imagination from its subjugation within' dominating images and discourse of what vulnerability, threat and risks are real, in order to 'create images which do political work of another kind'. Purposely situated early in the anthology, we see that his challenge to rethink the reality of threat is addressed in subsequent chapters, perhaps most alarmingly in the following chapter by Rebecca Hester.

Making explicit the security challenges of smart technologies, Hester suggests that smart bodies are seen as political and social sites, whereby biological sovereignty can be asserted in ethically harmful ways. She uses the term 'smart bodies' to refer to the widespread appeal of smart technologies and their purported ability for efficiency and optimisation while questioning the dangers of protological control. Inserting politics, economics and humanity into the quest for resilient bodies and resilient selves, Hester's chapter sounds the alarm on the unknown effects of resilience-building technologies and rationalities of governing life itself at the molecular level.

The chapter by Christopher Zebrowski and Daniel Sage tackles what has become an important theme emerging in neoliberal critique of resilience – There Is No Alternative. While recognising the legitimacy of many of the arguments about resilience as a neoliberal concept, they argue that we cannot resolve the collective challenges of ever increasing risks by critiques of resilience. They posit organisational studies as elemental to resilience machines. They use this lens to assess arguments about community versus society and explore the extent to which communities mobilise to counter the powers of neoliberalism as expressed through corporate and government actions. Drawing on the post-Katrina context, they suggest that elite managerialism is foundational to the use of resilience efforts as a means of control. Linking neoliberal economic ideology and managerialism represents an important advancement of critiques of resilience policy.

Raven Cretney's case study complements the previous chapter's analysis of post-disaster recovery. She challenges the view of resilience as a rigid body of governmentality. In her analysis of the responses to devastating earthquakes in Christchurch, New Zealand, she argues that post-disaster conditions create opportunities for resilience policy that are more progressive than the traditional neoliberal approaches that are often ascribed to recovery and resilience building. Although regressive approaches to recovery were evident in Christchurch after the disaster, space was created to construct new forms of citizens to relate to one another and their communities, creating what Nelson calls an 'ontology of potentiality' (Nelson, 2014, page 6). Rooted in evidence from Christchurch, Cretney examines how citizen engagement can intervene at the edges of neoliberal resilience initiatives, thereby promoting new norms, values and materiality in support of justice.

Lauren Rickards, Martin Mulligan and Wendy Steele argue that 'community' is the scale at which a resilience assemblage can best meet the needs of ordinary citizens, because communities can break away from the individualistic demands of neoliberal politics and permit self-organising responses to external stressors. They also consider community and resilience as 'conceptual twins' that align themselves along a spectrum of openness to change and create the potential for resisting resilience machines. This framing leads to a typology of interactions between community and resilience where the intersections of bound and unbound communities and resilience create opportunities for different responses to stressors. Where resilience and community aesthetics are bound, response conforms to maintenance of the status quo but, where both community and resilience are unbounded, the emergent properties might foster new relationships that can create creative and transformative change.

Kevin Grove and Jonathan Pugh offer a detailed account of resilience machines at work in the Caribbean. They note the divisionary paradox of resilience that has come to define two camps – those who believe in the potential of resilience to be used for social change and those who employ resilience to perpetuate staid political and economic power. Grove and Pugh deploy the concept of 'adaptation machines' to examine catastrophe insurance in Jamaica. They interrogate the depoliticisation of environmental governance and recentre the diffuse and affective dimensions of resilience machines in order to advocate for more ethico-aesthetic paradigms of resilience.

Peter Rogers' contribution focuses on the relationship between resilience strategies and implementations – on *how* resilience is done in the urban context rather than *what* it means. Rogers argues that attention should be paid to how resilience is acted out by resilience practitioners who can make or break resilience ideals depending on how they translate and implement resilience strategies. Acknowledging the significance of the action on the ground, Rogers turns his attention on strategies themselves. Two international urban policy frameworks, which position resilience as the core activity of governance, are analysed: HABITAT III: The New Urban Agenda and Rockefeller 100 Resilient Cities. Rogers contends that animating the power that practitioners have to mainstream resilience as an everyday practice allows us to view what sort of relationships are required to intervene into resilience machines. As Rogers suggests the potential for collaborations with resilience officers, the political costs of such bargaining remains unclear. The chapter closes with a call to consider civil society as a key aspect of resilience and as a technique of governance rather than its outcome.

Thilo Lang also emphasises the progressive potential of resilience. Using a critical urban and regional studies perspective, Lang discusses the limits and opportunities of the resilience machine. He highlights the challenges faced by rationalistic traditions of urban planning in times of uncertainty and complexity. Drawing on the evolutionary understanding of resilience, he frames cities as complex adaptive systems. His chapter puts forward a number of propositions that can be used as an analytical framework for future research. These include the need to: adopt a relational understanding of cities (or more broadly space); pay attention to the institutional structures and powerful resilience coalitions; move away from equilibrium thinking and a desire to bounce back; challenge the neoliberal instrumentalisation of resilience creating adaptive spaces that can accommodate changing demands of capital accumulations; and explore the progressive potential of the resilience machine in pursuit of justice and democracy.

Issues of justice and fairness are at the heart of Susan Fainstein's engagement with resiliency planning for New York City following Hurricane Sandy. She compares the ways in which resilience has been used at the mayoral level. She suggests that Mayor Bill de Blasio's plan failed to go beyond the rhetoric of equity and respond to the call for environmental justice in relation to the location of unwanted uses in the city. Fainstein argues that some practices of 'making room for water' in areas where the affected lands are home to low-income communities are examples of unjust adaptation planning. She looks to tax plans to demonstrate how the resilience machine converges with the urban growth machine in creating wealth for the elites. She concludes by highlighting the existence of counterweighing pressures from community organisations and environmental justice groups who are demanding an analysis of who loses and who gains.

Highlighting the hope for more progressive outcomes, Brendan Gleeson's chapter strikes a hopeful tone. Inspired by Hannah Arendt's work, Gleeson defines goodness in terms of human flourishing, enabled and enhanced by new urban imaginaries. He challenges current urban narratives, such as the mechanistic and technocratic discourses of urban resilience that normalise the apocalypse and advocate the status quo. He calls for an urban imaginary that fuses reason with faith. Gleeson's question about the good city emphasises the role of faith and hope especially in the condition of radical uncertainty and the logical impossibility of knowing the unknown. Paraphrasing Eagleton (2015), Gleeson argues that 'resilience without hope' is nothing more than accepting that capitalism faces dissolution. For Gleeson, resilience is about the continuation of hope and the endurance of the cycle of birth, death and rebirth.

Closing the volume, Timothy Luke assesses resilience as a 'restoration engine' rooted in conservative impulses that support agendas of stability. As a conceptual, ethical and operational apparatus for restoration, the resilience machine demands that individuals, groups and places should confront disruptive changes or systemic challenges to their existence by restoring themselves to some illusive status quo ante; that its function is to prevent progressive or positive changes. Luke argues that exploring these dynamics in contemporary corporate and government discourses is essential for understanding the instrumentalist uses of resilience as a means of managing or mitigating change to the point of preventing it. Importantly, Luke reminds us that when growth stops, resilience is pushed forth to regain it. Problematising the prevailing notion of bounce-back-ability in the face of increasingly strong and frequent storms, Luke contends that resilience is often unable to change the conditions that are productive of disaster.

Animating Questions, Themes and the Future of Resilience Machines

When we began the project, we posed a number of questions for our contributors to consider as they composed their chapters. It became obvious through our workshop discussions that a number of strong themes were binding the project together as a cohesive look into the way resilience is operationalised as a machine. Questions such as the ones mentioned below have not only informed our project, but also might be helpful for others to understand the relationality and assemblages of resilience machines: What values are inscribed and prescribed in the name of resilience? What actors and institutions deploy instrumentalist forms of resilience? Moreover, what social and political effects emanate from resilience initiatives? These questions have also raised other issues related to resilience as we have moved through this project and witnessed new iterations of resilience coming online.

We recognise that resilience critique has become en vogue. Still, we believe that bringing attention to the pervasiveness of resilience as a way to manage the uncertainties of community relations, resource flows in the city and the vulnerabilities of human bodies, ecosystems and life itself is worthwhile, given the widespread deployment of unexamined 'resilience' policies and practices. In *Liquid Modernity*, Zygmunt Bauman articulates our desire to understand our fate in a constantly changing world. He notes: 'To understand one's fate is to know the complex network of causes that brought about that fate and its difference from destiny' (Bauman, 2000, page 212). The knowing and security that resilience offers in the face of life's uncertainties is liquid. Nevertheless, as we cope and continue to work in the world, we struggle to 'know how the world works' (ibid.). In this spirit, we draw into focus the agents and agencies whose work simultaneously produce vulnerability and respond with resilience as a self-legitimating ontology. Identifying the networks of discourses, agents and assemblages demonstrates the production of resilient subjectivity and, highlighting the networked ways in which resilience and uncertainty cleave together, makes obvious how resilience-building strategies allow the reification and circulation of resilience as an uncontested good in the public domain. This contradiction is central to the work on *The Resilience Machine* because, as François Debrix (2018, page 260) reminds us, vulnerability requires 'a corresponding modality of governance/governmentality best suited (or so we are told) to manage' the precarious conditions of contemporary life. 'Thus, for example, resilient life calls for, justifies, and makes effective operations, technologies, and strategies of resilience.'

For most of us, resilience is *not* recognisably integrated into our everyday lives. While we may regularly consider how to cope with ordinary stresses of work, relationships, health and finances, the functioning of the resilience machines and their power in shaping our daily existence remain largely occluded. In part, this anthology was originally envisioned as a way to shed light on the processes, actors, institutions and influences that are embedding resilience into life itself; from prosaic acts like walking and moving in cities (Rogers, Chapter 8), to technological interventions into health outcomes (Hester, Chapter 3), disaster management and regional planning (Lang, Chapter 9). The invisibility of resilience does not automatically constitute nebulous and nefarious business operations, or collusion between governments and industries to responsibilise individual subjects or particular populations. Rather, resilience is manifest in a multitude of settings and in numerous ways. The values underpinning resilience strategies, for example, can vary across domains, as can the effects of these strategies.

Locating the Resilience Machine

Despite the fluidity of resilience machines and their operations, we can see their effects in a number of contexts. The abstract nature of resilience does not mean that it cannot be located. Indeed, through the volume, one of the key referents in resilience discourse is place, or to be more precise, cities and communities. This is also encapsulated in the growing academic and policy literature on urban resilience as well as the avalanche of programmes by non-governmental organisations such as Rockefeller Foundation's 100 Resilient Cities initiative, and the production of multiple city resilience indices (Arup, 2014). The city as an assembly of assemblages with complex and uncertain socio-spatial relations is a fertile ground for the construction and operation of the resilience machine. A distinct characteristic of urban resilience narratives is the reference to digital technologies, big data and algorithms. Cities across the world are capturing vast quantities of data, largely to the benefit of multinational corporations such as IBM and Microsoft, in order to construct 'automated resilience' (Stark, 2014). The highly instrumented, all-seeing city promises a fully automated, 'smart' resilience machine. The promise of automation, however, ought to come with a warning. While it is sold as the next frontier of resilience building, it can have dire distributional and democratic consequences (see Knight, 2017). Automated failures are destined to occur, leading to calls for more resilience, rather than inquiring into the logics that have created that demand in the first place. Questions of accountability, therefore, infuse the contributions to *The Resilience Machine*.

Resilience as a Will to Knowledge/Control

Attempts to engage with resilience as a form of knowledge and control is examined throughout the chapters. The fluidity of resilience and its production, circulation, reification and materialisation within a system does not mean that attempts to engage with the intentionality of resilience is impossible, nor that such analyses have to avoid targeting knowledge and control. In tracing the anatomy of the resilience machine Chapter 1 makes it clear that there are inflection points at which power/knowledge hierarchies are exercised. This is further articulated by Grove and Pugh (Chapter 7), Lang (Chapter 9) and Fainstein (Chapter 10). Reflecting on the power/knowledge dimensions that are linked to the material effects of resilience is hinged to political ideology. As Sheila Jasnoff notes: 'ideology is generally seen as entrenched and immovable. Ideology also lacks the imagination's properties of reaching and striving toward possible futures, and ideology has not typically been analyzed as being encoded in material technologies' (Jasanoff, 2015, page 29). How, then, can we envision alternative ontologies of resilient subjectivity if assemblages are fundamentally oriented to a metastatic imperative of economic growth as an ideology? If resilience machines are super intelligent, replicating and embedding themselves into the functions of everyday life, would it be possible to slow, halt or avert them, even if we intended to?

Resisting Resilience Machines

Almost all of the chapters in this volume provide examples of how the unfolding politics of the resilience machine can open up spaces for resistance and the emergence of alternative ontologies. The relational aspect of resistance to power recognises the force of assemblages, but confronts the idea that there is a deterministic aspect to resilience. Rather, it is the manifestation of resilience in context 'at which force is translated into power' (Massumi, 1992, page 31). This is particularly articulated by Cretney (Chapter 5) and Zebrowski and Sage (Chapter 4). Their case studies evidence the ever present 'possibility space' which, as we discuss in Chapter 1, is an idea invoked by both assemblage and complexity theory. The former highlights the contingency of assemblage and the existence of the potential for it to be morphed into something else at any point in time and space. The latter stresses the opening up of 'windows of opportunity' when complex adaptive systems collapse. Both reject the linear and predictable cause and effect relationship and instead emphasise the contingent and emergent nature of the resilience machine. While acknowledging that the resilience machine can open up space for rupture and dissent, many contributors warn that bottom-up initiatives can be easily morphed by rigid metrics of responsibilities and deliveries and become co-opted into the dominant neoliberal strategies.

In resisting the instrumentalisation of resilience, many of our contributors unravel the interwoven nature of the material and expressive assemblages of the resilience machine. Identifying these characteristics as inherent to resilience, they expose the normative dimensions of resilience that require rethinking for a more equitable, just and sustainable form of living to emerge through resilience initiatives. This is particularly evident in the discussions of post-disaster community and urban resilience. What separates the chapters is a matter of degree rather than an either/or perspective on the nature of the resilience machine and its ramifications for people and place. Some put the emphasis on its progressive opportunities and the hope for transformation (Gleeson, Chapter 11), others on its regressive consequences and need to dismantle assemblages of resilience (Luke, Chapter 12). We believe that both are ever present and it is the ongoing struggle between them that carries the potential to stop the depoliticising tendencies of the resilience machine.

Holding Out Hope for Resilience?

For many, *The Resilience Machine* will create discomfort. Practitioners of resilience policy and advocates of cultivating resilient subjectivities are systematically entangled in a governance strategy designed to cope with vulnerability, as are those who offer critiques of these strategies. Nevertheless, interrogating the conditions that produce (what many believe to be) resilience as a 'social good' and to which many dedicate their daily work can be jarring. We were confronted with this unease at a recent American Association of Geographers conference where we presented the initial ideas that led to this anthology. However, we see this unease as a threshold whose creative potential ought to be captured. As Brown and Strega (2005, page 128) suggest: 'What is most exciting and creative about thresholds as passageways are the possibilities that are produced by letting go of destinies and expectations, by learning to live with and through uncertainty'.

This experience encouraged us to continue the project. While the unveiling of sometimes dangerous effects of resilience machines sparked unease, the project is not intentioned to cast judgement on individuals and their good will. Rather, it is an effort to put forward a critique of the systemic production of the need for resilience. For the machine to work, the system must be vital and the agents must be operational, but that does not mean that it is free from resistance. So, should we hold out hope for resilience? Do creative and critical potentials still rest within resilience discourses, policies and practices? If so, how might we imagine it?

The preponderance of resilience machines means that a unified framework to understand these assemblages is neither desirable nor possible. Confounding the struggle for resilience that allows life to flourish, we are encountering mass extinction. Reflecting on the Anthropocene, Elizabeth Povinelli urges us to view the fascination with the death of ourselves, fellow beings and potentially the planet itself as a form of 'nonlife' that 'holds, or should hold for us, radical potential' (Povinelli, 2016, page 176). But, what might this radical potential produce, and are we capable of imagining a future that operates differently from our present world? For Donna Haraway (2016, page 86), the answers lie in 'inventive, sympoietic collaborations' as we learn to live/die on a damaged planet. This optimistic view is echoed by Stephanie Wakefield (2017, page 6), who draws together the narratives of resilience and the Anthropocene, urging us to make the politically and empirically necessary step of 'inhabiting the back loop' of adaptive resilience (see Chapter 1) as we learn to live with vulnerability and planetary boundaries. Riffing on the double loop of adaptive resilience, she reflects on the conditions of our contemporary world as Anthropocenic, and encourages us to 'push resilience thinking's own boundaries, especially as it pertains the deep potential for transformation at the heart of its foundational heuristic' (Wakefield, 2017, page 6).

Yet, governing planetary boundaries through such a transformational lens still requires the everyday and mundane stressors of living in the Anthropocene to be addressed. Moreover, we must remain aware that the discourse of the Anthropocene can reinforce an anthropocentric view of the world that affirms 'the centrality of man – as both causal force and subject of concern' (Crist, 2013; see also Davoudi, 2014). Thus, 'the very concept of the Anthropocene crystallizes human dominion' (Crist, 2013) and limits our ability to have a critical gaze. This cannot be taken for granted and we must take caution not to consistently refashion efforts to secure against risk, modulating life as a technology in need of governance and control. As discourses of resilience proliferate they sculpt our expectation for life. Perhaps within this multiplicity potential new ways of thinking about resilience still reside.

Reflecting on the provocations presented in this anthology, we propose that *if* there is critical potential left within resilience, it needs to move beyond mechanisms of adaptation and into imagining what resilient life could look like in a more just, equitable and democratic global society. This is not about bouncing back, bouncing to the side or bouncing forward in the same systems that have created and exacerbated vulnerabilities. It requires a new ontology. Some possible ways forward are intimated in some of the contributing chapters but this deserves further attention. Such a version of resilience would not bolster systems that produce uneven geographies, or imagine futurities without vulnerabilities. Rather, it would dismantle the most resilient systems of exploitation that have produced suffering for many human and more-than-human lives. A future like this is difficult to imagine given present political, environmental and social challenges. Nevertheless, the liberating potential of resilience may rest in accepting the inherent vulnerability of life itself while simultaneously rejecting the powers that create and exploit and disproportionate vulnerabilities for certain people, places and ecosystems. As Nevzat Soguk (1999) reminds us, a predetermined course of action for resistance does not have to be set forth, nor does critical potential need to be reduced to specific tools or questions. What is revealed through a critical interrogation of resilience is what matters most; through these revelations the opportunity to contest the very assumptions on which resilience is built might be rethought. We must inquire into what appears to be necessary, universal and above questioning. Tethering the progressive and regressive effects of resilience means that we must remain engaged with resilient life as a heterotopia because the assemblages that enable particular lives and livelihoods to flourish also makes invisible vulnerability, thereby obscuring the sources that produce precarity. Exposing the operations, agents, institutions and effects of resilient systems of exploitation allows us to refocus on responsibility rather than centring on resilience – a task for scholars who are genuinely interested in imagining a world that can be otherwise.

References

Aranda, K., Zeeman, L., Scholes, J. and Morales, A.S.M. (2012). 'The resilient subject: exploring subjectivity, identity and the body in narratives of resilience'. *Health*, 16(5): 548–563.
Arup (2014). *City resilience framework, city resilience index*. New York City: Rockefeller Foundation.
Åsberg, C. (2014). 'Resilience is cyborg: feminist clues to a post-disciplinary environmental humanities of critique and creativity'. *Resilience: Journal of the Environmental Humanities*, 1 (1): 5–7.
Bauman, Z. (2000). *Liquid modernity*. Cambridge: Polity.
Brown, L. and Strega, S. (eds) (2005). *Research as resistance: critical, indigenous and anti-oppressive approaches*. Toronto: Canadian Scholars' Press.
Chandler, D. (2014). *Resilience: the governance of complexity*. Abingdon: Routledge.
Crist, E. (2013). 'On the poverty of our nomenclature'. *Environmental Humanities*, 3(1): 129–147.
Davoudi, S. (2014). 'Climate change, securitisation of nature, and resilient urbanism'. *Environment and Planning C: Government and Policy*, 32(2): 360–375.
Debrix, F. (2018). 'End piece: dealing with disastrous life', in J.L. Lawrence and S.M. Wiebe (eds), *Biopolitical Disaster*. Abingdon: Routledge, pp. 257–263.
Eagleton, T. (2015). *Hope without optimism*. Charlottesville, VA: University of Virginia Press.
Haraway, D.J. (2016). *Staying with the trouble: making kin in the Chthulucene*. Durham, NC: Duke University Press.
Jasanoff, S. (2015). 'Future imperfect: science, technology, and the imaginations of modernity', in S. Jasanoff and S.-H. Kim (eds), *Dreamscapes of modernity: sociotechnical imaginaries and the fabrication of power*. Chicago, IL: University of Chicago Press, pp. 1–47.
Knight, W. (2017). 'The dark secret at the heart of AI'. *Technology Review*, 120(3): 54–61.
Massumi, B. (1992). *A user's guide to capitalism and schizophrenia: deviations from Deleuze and Guattari*. Cambridge, MA: MIT Press.
Molotch, H. (1976). 'The city as a growth machine: toward a political economy of place'. *American Journal of Sociology*, 82(2): 309–332.
Nelson, S.H. (2014). 'Resilience and the neoliberal counter-revolution: from ecologies of control to production of the common'. *Resilience*, 2(1): 1–17.
Povinelli, E.A. (2016). *Geontologies: A requiem to late liberalism*. Durham, NC: Duke University Press.
Soguk, N. (1999). *States and strangers: refugees and displacements of statecraft* (Vol. 11). Minneapolis, MN: University of Minnesota Press.
Stark, D. (2014). 'On resilience'. *Social Science*, 3: 60–70.
Wakefield, S. (2017). 'Inhabiting the Anthropocene back loop'. *Resilience*, 5(3): 1–18.
Walker, B. and Salt, D. (2012). *Resilience thinking: sustaining ecosystems and people in a changing world*. Washington, DC: Island Press.

1
ANATOMY OF THE RESILIENCE MACHINE

Simin Davoudi, Jennifer Lawrence and Jim Bohland

Introduction

> Forget sustainability. It's about resilience.
>
> (Zolli, 2012)

The above *New York Times* headline heralded resilience as the dominant discourse of our time. But, does the ushering in of resilience mean that the social, economic and environmental pillars of sustainability have crumbled? That the sustainability agenda has not been successful in addressing the seemingly intractable economic, social and environmental crises which contemporary society is grappling with? Global inequality, for example, has reached an all-time high with half of the global wealth owned by just eight men, and the other half unevenly divided between 3.6 billion people (Oxfam International, 2017). Reporting from the 2017 annual meeting, the World Economic Forum (2017, page 39) indicated, 'the time to act (on equality) is now' if we are to stem the 'deterioration of government finances and the exacerbation of social unrest' that has come to dominate our modern world. The call for fundamental reform in the capitalist world system, though, has yet to have an impact. Rather, the insecurity embedded within the grotesque state of global economic disproportionality has brought about a dangerous rise in far-right populism.

This turn to the far right is steeped in security rhetoric, whereby human rights are seen as a complication in the way of protection against a range of 'perceived threats and evils' perpetrated by immigrants, terrorist organisations and minorities (Roth, 2017). Such politics seize upon discontent with the status quo, and enables the accumulation of power and wealth while simultaneously eroding elemental features of democratic engagement. Compounding social and economic turmoil are the environmental pressures that mark further ramifications for disadvantaged communities. It is widely understood that those least responsible for the production of climate change will suffer disproportionately from its effects. Moreover, the conflict and ongoing exodus of refugees worldwide demonstrate that there are massive stakes for geopolitical security that attend this tripartite crisis. Indeed, many militaries are preparing for refugee crises of 'unimaginable scale' that are expected as mass migration coinciding with climate change increases (Carrington, 2016).

In response to the complexities and uncertainties of these interconnected crises the concept of resilience has found an upsurge in both academic writings and public policies. In the last three decades, it has become prominent across natural, physical and social sciences, and engaged scholars from a wide range of disciplines including engineering, economic geography, psychology, urban studies, political ecology, urban planning, public administration and disaster risk management, among others. Moreover, instrumentalist modes of resilience which often urge cities, citizens and communities to adapt to varying forms of distress have effectively colonised multiple areas of public policy and decision making. Today, a growing number of think tanks, philanthropic organisations, government institutions, non-governmental organisations and corporate entities have made resilience thinking and resilience building a top priority. Notable international examples include the campaign launched by the United Nations Office for Disaster Risk Reduction in 2012 on *How to Make Cities More Resilient*, the World Bank's 2012 guidance on *Building Urban Resilience in East Asia*, the Rockefeller Foundation's much publicised campaign for *100 Resilient Cities* and the *Resilience Alliance* which is populated by large corporations seeking new markets for their products. As a result, there are now multitudes of toolkits and guidelines on how to achieve resilience, as well as indicators and metrics about how to measure and monitor the resilience of cities, communities, individuals and ecosystems. In short, resilience has become a concept travelling far and fast.

However, the widespread currency of resilience is by no means a sign of common understanding of the concept, and the pervasiveness of resilience discourses does not indicate deep engagement with institutional arrangements – neither those that have precipitated the need for resilient subjectivities, nor those that benefit from the need for resilient subjectivities. In fact, as a result of the widespread deployment of resilience discourses, the concept risks being an empty signifier which can be filled with multiple meanings and serve conflicting interests. While some advocates indicate encouraging possibilities for resilience as a new way of thinking about and governing risk and uncertainty, others highlight the danger of resilience becoming yet another carrier of neoliberal practices with negative implications for social justice and democracy. Despite, or maybe because of the discursive ambiguity of resilience, the concept appears to have offered a common, albeit contingent and temporary, platform which binds and unites otherwise pluralistic interests. We see a 'resilience machine' in the making.

Multiple imaginaries and technologies of 'resilience' have been set forth as a response to profound, and interconnected, economic, social and environmental challenges. There exist tensions between the powerful, and often dangerous, effects of resilience discourses and the possibility of resilience to offer an alternative political orientation that fundamentally addresses the conditions that necessitate resilient individuals, communities and societies. In the wake of acute crises, cities, citizens and communities are often praised for their resilience and for their ability to press on in the face of devastation, without questioning the political, economic and social conditions that have created the need for resiliency in the first place. We would argue that narratives of resilience can mask the structural conditions that perpetuate the need for resiliency in the face of exploitation. From the outset, we acknowledge that our take on resilience cannot be politically neutral; ours is based on an egalitarian outlook, a democratic spirit and a pluralist ambition. In the following account, we trace the genealogies of the two concepts, 'machine' and 'resilience', that define the tenet of this chapter (and the book) and discuss how the resilience machine may be better understood as an 'assemblage'. The final section concludes the chapter.

The Genealogy of the Machine

The idea of 'machine' is ubiquitous and much like resilience has been a travelling concept. The term itself comes from French via Latin, from the Doric Greek makhana/mēkhos ('contrivance'). As a noun, its dictionary definitions include: 'An apparatus using mechanical power and having several parts, each with a definite function and together performing a particular task. Any device that transmits a force or directs its application. An efficient and well-organised group of powerful people. A person who acts with the mechanical efficiency of a machine' (Oxford Dictionary, 2017).

These definitions show the range of attributes described by the notion of 'machine', from efficiency to mechanistic approaches and to powerful forces. But, its high currency as a metaphor lies in its 16th-century origin and its association with the Scientific Revolution. This was the time of a momentous shift from the medieval view of the world as *organic* to the modern view of the world as *a machine*. Epistemologically, the 'organic' world combined spiritual faith with reason while the mechanistic world was based on pure reason. Normatively, the aim of the organic worldview was to understand the meaning of existence and achieve coherence in the human–nature relationship, while the aim of the mechanistic worldview was to predict the laws of nature and control its consequences (Capra, 1982; Davoudi, 2012a).

It is, therefore, not surprising that the origin of the term 'machine' can be traced back only to the 16th century when the towering figures of the Scientific Revolution (such as Nicolas Copernicus, Galileo, Francis Bacon, Rene Descartes and Isaac Newton) made the view of the world as a machine, governed by hidden mathematical laws, into the dominant paradigm of science and a powerful metaphor of the modern time. In this mechanistic, clockwork universe, there was little room for anything else but wheels and springs. Descartes, an enthusiastic advocate, said it all: 'I consider human body as a machine ... My thought ... compares a sick man and an ill-made clock with my idea of a healthy man and a well-made clock' (quoted in Capra, 1982, page 54). His principles along with those of Newtonian physics were carried through to the newly created social sciences in the 18th century. They influenced philosophers such as John Locke whose atomistic view of society provided the moral order of modern social imaginaries (Taylor, 2004) based on values such as individualism, free market, property rights and representative democracy.

In the 19th century a number of scientific advances began to challenge the view of the world as a machine. One of the most influential was the idea of evolution (Capra, 1982). Comprehensively developed by Charles Darwin, evolution theory transformed the field of biology and replaced its long-held assumption about 'a great chain of being' made of a rigid hierarchy of biological species with the idea of random mutation and natural selection. Evolution theory also entered the field of physics and led to the formulation of complexity science which is the foundation of the evolutionary resilience, as discussed below.

However, a key difference exists in the way evolution was applied in biology and physics. In physics, the study of thermodynamics led to two fundamental laws of physics: the conservation of energy and the dissipation of energy. The former suggests that energy is never lost; it just changes from one form to another. The latter suggests that although total energy is always conserved, useful energy dissipates in the process of conversion and this leads to increasing *disorder* in the system. The term entropy (energy and *tropos*, the Greek word for evolution) is used to define a quantity that can measure such disorder (Capra, 1982; Hornborg, 2006). Resilience might be interpreted as a way of controlling disorder. By contrast, in biology evolution implies a movement towards complexity and *order*, as simpler forms evolve into complex systems. Here, resilience might be seen as creating alternative order.

Together, these theories changed the simplistic and mechanistic assumptions about the world as a machine. Throughout the 19th century, evolutionary thinking shaped the work of political philosophers such as Hegel, for whom the key philosophical question was not the problem of being but becoming. However, these scientific advances have not reduced the metaphoric power of 'machine' and its frequent use in a myriad of contexts. One example is the Scottish polymathic planner and botanist, Sir Patrick Geddes, who became a demonstrator in practical physiology at the University College London. In 1879, while travelling to Mexico to collect biological specimens, he suffered from poor eyesight. That led him to devise the idea of 'thinking machines', a series of visual methods for presenting and connecting facts and ideas to better understand the evolution of cities (Boardman, 1978).

A more recent example is the highly acclaimed book by Alf Hornborg (2006), *The power of the machine*. Using the term 'machine' to describe modern technology, he shows how it engenders asymmetrical distribution of risks and resources between distant societies and ecosystems, leading to an increasingly polarised world. He offers a theoretical analysis of how technical fetishism masks the inequalities between societies and ecosystems across the globe. Although an anthropologist by background, Hornborg draws on the second Law of Thermodynamics to reconceptualise industrial technology and unmask the exploitative nature of technological development which, according to him, should be understood as 'a cultural (i.e. simultaneously materials-social-symbolic) institution regulating the distribution of resources on a global scale' (page 126) and as 'a strategy of elite capacitation' (page 153) because, 'any local accretion of industrial technology can be reproduced only by accelerating the pace of dissipation elsewhere' (page 124).

Hornborg's work is a testimony to the metaphoric power of machine both in describing technological development and in perpetuating the dominant worldview which he calls 'a Western "world structure"' (Hornborg, 2006, page 128). As he suggests, 'the isolation of form from process and of parts from whole is an aspect of the proclivity to objectify that has been so characteristic of Western thought. It produces an atomistic, reductionist worldview supremely adequate for technical manipulation but hardly for holistic understanding' (page 119). These resonate with the criticisms of the resilience machine and its tendency to atomise and individualise, as discussed later in this chapter.

The machine metaphor has also been used in urban politics, notably in the description of the urban political institutions of the late 19th and early 20th centuries in the United States. A classic example is Clifford's (1975) detailed account of the rise of a 'political machine' in the rapidly growing industrial cities and in response to massive immigration from Europe and rural areas. He used the metaphor of the machine as a powerful force to describe the hierarchical structures of urban governance, organised around a single authoritarian figure and operated through extensive networks of patronage and subordination. These 'political machines' ruled the city by controlling the distribution of resources and economic opportunities, and determining who got what, how and where (Lasswell, 1936).

A final example of the use of the machine metaphor, one which provides the inspiration for the title of this book and is closer to our use of the term 'machine' in this chapter, is Harvey Molotch's seminal paper, 'The city as a growth machine' (1976). At the heart of his thesis, which was elaborated in his later work with John Logan on 'urban growth machine' (Logan and Molotch, 1987), is that urban politics is essentially driven by a coalition of land-based elites, not political machines, who not only exploit the economic potential of rising land values in localities to accumulate wealth, but also actively manipulate the political, regulatory and financial systems that govern land development to secure the precondition for growth and profit from it. The manipulation of these material preconditions complements ideologies and discourses that strive to convince people of the importance of growth for their jobs, pensions, health and wellbeing. So, the anatomy of growth machines (like the resilience machine) is made of not only individual actors and institutions, but also ideologies and discourses. Together they generate 'the desire for growth (which) provides the key operative motivation toward consensus for members of politically mobilized elites' (Molotch 1976, page 310) as well as for the public at large.

The growth machine thesis was a significant contribution to urban political theories and provided one of the fundamental insights into 'the specific mechanisms through which space was produced and life chances thereby distributed' (Molotch, 1999, page 247), particularly in 20th-century American cities. However, the 21st-century city is no longer a mere growth machine, even though it is highly driven by it. A review by Lauermann (2016) shows that the financial crisis and austerity politics have led to diverse forms of 'municipal statecraft' engaged in a myriad of experimentations with urban policy. These have diversified the local elites' focus on growth and introduced more diffused forms of policy boosterism (McCann, 2013), 'growth machine diaspora' (Surborg et al., 2008, page 342), 'de-growth machines' (Schindler, 2016), the 'debt machine' (Peck, 2016) and we would argue, the 'resilience machine'.

It is argued that urban entrepreneurialism is repurposed and applied to multiple agendas and diversified investment and policy portfolios which involve not only speculation, but also *experimentation*. This is reflected in the growing number of 'urban labs' that are experimenting with 'smart city', 'eco city', 'world city', 'circular economy city' and indeed 'resilient city'. Here, policies are evaluated on the basis of not only their economic returns, but also their self-defined set of metrics, such as resilience indicators. These provide municipalities with multiple ways of claiming local political legitimacy and global acclaim. None of these suggest that the move is progressive or that it replaces the obsession with growth. Indeed, as contributions to this book show, although many of these urban experimentations may broaden the local political agenda, they nevertheless retain their elitist tendencies because, as Smith et al. (2014, page 24) suggest, 'it is very hard to say no to rich, powerful actors, particularly when their preferred policy appropriations are framed as privately financed, and thus publically costless, experiments'. Growth continues to be the touchstone of urban policies and resilience is often called upon as a remedy to stop the slowing down of urban growth. Furthermore, the increasingly international dimension of inter-urban competition of recent years has led to machine-like coalitions geared to property-led regeneration schemes with one striking difference. Today's coalitions attract transnational corporations and draw on international capital and markets while engaging locally in tactical politics around land-use regulation, urban policy and planning and also the rhetoric of resilience.

The above discussions show that the metaphor of machine is apt in capturing the advocacy of resilience as a taken-for-granted public good, but our understanding of the anatomy of the machine and the way it works differs from Molotch's thesis in important ways. First, the urban growth machine is based on an agency-centred perspective which puts the emphasis on actors' observed action, rather than the social relations that give rise to their action. We argue that while the resilience machine is activated and mobilised by the agency of actors, their actions are shaped by the institutional structures and the social milieu within which they operate. Second, the growth machine operates in relation to just one form of urban politics, i.e. land and its commodification. The resilience machine is pervasive. It is active in multiple forms of urban politics ranging from citizenship, to community relations and access to welfare provisions. Third, Molotch's thesis is uniscalar and relates primarily to late 20th-century American cities. The resilience machine operates at multiple interconnected scales from international to neighbourhood and even individual levels. Based on a relational, rather than absolute, understanding of scale, we argue that the resilience machine plays an active role in the social construction of scale itself. Fourth, like Molotch, we see language as a critical intellectual currency in the co-option of actors and interests into the machine's operation. But, unlike Molotch, we consider language as well as other forms of significations (such as images, codes, algorithms and performances), not as simple communication devices for convincing the public about the benefits of growth for their wellbeing, but as performative acts aimed at normalising and essentialising specific forms of resilience practices into everyday life. Through various forms of significations, citizens, cities and communities are motivated to board the machine. Fifth, and finally, Molotch considered the feelings and concerns of 'residents' for localities as a 'counter-coalition' force that fights against self-serving, profit-seeking developers and growth entrepreneurs. We argue that resisting the resilience machine and its subjectification intents is not exclusive to a particular group of people based in a particular locality. It is diffused across time, space and social groups and is embedded in the power relations of resilience politics.

Our different and expanded understanding of the resilience machine resonates with the Deleuzian concept of *assemblage*, and our understanding of its politics follows Connolly's notion of the 'resonance machine', both of which are discussed in the next two sections.

Resilience Machine as Assemblage

The concept of 'assemblage' has served multiple purposes in contemporary social theories. It is often used to simply describe the coming together of heterogeneous elements in a city, an institution or a piece of art. Its significance lies in its theoretical insights developed through the work of Gilles Deleuze and Felix Guattari (1987). Translated from the French *agencement* to the English 'assemblage' by Brian Massumi, the assemblage theory strives to avoid reification, reductionism and essentialism, all of which has added to its growing appeal across the social sciences, notably the 'actor network theory' developed by Bruno Latour and Michel Callon. The broad applicability of the work of Deleuze and Guattari has enabled it to be used as 'a "tool box" – as a collection of machinic concepts that can be plugged into other machines or concepts and made to work' (Malins, 2004, page 84).

According to Manuel DeLanda (2016, page 1), who has produced the most authoritative account of the assemblage theory, the original French word '*agencement*' refers to 'the action of matching or fitting together a set of components (*agencer*), as well as the result of such an action: an ensemble of parts that mesh together well'. The English translation to 'assemblage' captures only the second meaning, giving the wrong impression that the concept is about the product not the process. In addition to translational imprecision, Deleuze's prose can be highly allusive as it is peppered with terminological innovations that change from one work to the other. Assemblage has not been immune to such innovations and has received several different definitions in his works. We believe the following definition encompasses some of the key features of his assemblage theory: 'What is an assemblage? It is multiplicity which is made of many heterogeneous terms and which establishes liaison, relations between them … Thus, the assemblage's only unity is that of co-functioning … It is never filiations which are important, but alliances' (Deleuze and Parnet, 2002, page 69).

Unpacking this definition, an assemblage is an emergent social whole produced through the interactions of diverse and heterogeneous forms of human and non-human (objects, events, signs and utterances) agencies. The whole is emergent in the sense that its qualities cannot be 'analytically tractable from the attributes of internal components' (Manson, 2001, page 410). Latour provides a useful description of how an assemblage is made from the ground up. 'Macro no longer describes a wider or a larger site in which the micro would be embedded like some Russian matryoshka doll, but another equally local, equally micro place, which is connected to many others through some medium transporting specific types of traces' (Latour, 2005, page 176).

The component parts of an assemblage are often in a relation of exteriority to the whole which enables them to retain their autonomy, detach themselves from the whole and plug themselves into another whole. Concepts of *emergence* and *exteriority* imply that although a social whole is more than the sum of its individual parts, it does not totalise them (DeLanda 2016). However, once an assemblage is produced, it acts as a source of opportunities and limitations for its constituent parts (Dittmer, 2014). Inherent in this understanding of assemblage is the existence of power relations which is often overlooked in the application of assemblage theory. Consistency and coherence between the component parts are emergent properties that may or may not arise from an assemblage. The term consistency is used not in the *logical* sense of being opposed to contradiction, but rather to describe the hanging together of heterogeneous parts. As Tampio puts it: 'The brilliance of the concept of assemblages is that it describes an entity that has both consistency and fuzzy borders … [it] has some coherence in what it says and what it does, but it continually dissolves and morphs into something new' (Tampio, 2009, page 394).

The interactions between the parts and their effects are *indeterminate* and *complex* and this makes it impossible to predict the behaviour of the whole even when extensive knowledge of its parts is available. Furthermore, assemblages can multiply in every direction at various speeds, timeframes and scales, much like a complex adaptive system in resilience thinking. Deleuze and Guattari distinguish between 'machinic assemblages' and 'collective assemblages'. DeLanda (2016, page 7) uses this to distinguish between material and expressive components while acknowledging that they are interrelated. In the context of resilience, machinic assemblages are materials such as physical flood defences (walls, barriers, etc.) while collective assemblages are expressions found in, for example, flood risk policies and programmes or symbols such as flood warning sirens. There are clear conceptual proximities between Deleuze and Guattari's theory of assemblage as a way of thinking about the social world and its multilevel ontology, and the complexity theory which underpins the evolutionary understanding of resilience, as elaborated in the next section. One indication of their proximity is Thrift's (1999, page 33) definition of complexity as 'the idea of a science of holistic emergent order'.

Overall, assemblage theory can contribute to our understanding of how the resilience machine is produced, maintained and subjected to change in multiple ways (McFarlane, 2011, page 208). As a *description*, assemblage implies that the resilience machine is produced through the ensemble of actual and potential relations between its heterogeneous components. As a *concept*, assemblage highlights the role of political agency in both enacting and rejecting the resilience machine. As an *imaginary*, assemblage invokes the taken-for-granted and invisible assumptions that are embedded in and enabled by the resilience machine. And, as a focus on *process*, assemblage highlights the unfolding set of uneven practices and relations that make, remake and resist the resilience machine. Assemblage theory also provides a set of *analytical tools* for deconstructing the anatomy of the resilience machine and its rationalities, imaginaries and practices, as demonstrated by a number of contributors to this book.

Genealogy of Resilience

The term resilience has a long history with multiple origins. Etymologically, resilience comes from the Latin word *resilire*, commonly known as rebound or spring back. According to Alexander (2013), the first known scientific use of the term in English is associated with Sir Frances Bacon, the father of modern scientific method, who referred to it in his 17th-century writings on natural history. In the mid-19th century, resilience was used to signify the ability to *recover* from adversity. At the same time, the term resiliency was used by Americans in reaction to the Shimoda earthquake in 1854 to refer to the ability to *withstand* stress. In mechanics, it first appeared in the work of the Scottish engineer, William Rankine, in 1858. He considered a resilient steel beam as both rigid (*resist* forces) and ductile (*absorb* forces); two characteristics that continue to define the mechanistic use of the term in the present time. Its adoption in social sciences goes back to the 1950s (Bloch et al., 1956) and especially the research by Norman Garmezy into children's developmental psychology (Garmezy et al., 1984).

However, the current popularity of resilience and its widespread use is largely due to the work of Crawford Stanley Holling, a Canadian ecologist. In his seminal paper in 1973, he challenged the *engineering* understanding of resilience which considers 'the resistance to disturbance and the speed by which the system bounces back as the measures of resilience' (Davoudi, 2012b, page 300). Instead, he put the emphasis on adaptation and suggested that, 'Resilience ... is a measure of the ability of these systems to absorb changes ... and still persist' (Holling, 1973, page 17). In this *ecological* understanding, resilience is defined as 'the *magnitude* of the disturbance that can be absorbed before the system changes its structure' (Holling, 1996, page 33). Here, the measure of resilience is 'not just *how long* it takes for the system to bounce back, but also *how much* disturbance it can take and stays within critical thresholds' (Davoudi, 2012b, page 300). Holling's main contribution was his adoption of complexity theory and multiple equilibria in system dynamics instead of accepting the existence of a single, steady equilibrium.

Following Holling's contribution, resilience has appeared in a growing number of academic papers, especially in environmental studies journals. Furthermore, its use has been morphed from the natural and physical sciences to the social sciences and has intensified since the 1990s, as is evident from the 400 percent rise in the number of papers using resilience in their topic in the *Social Science Citation Index* between 1997 and 2007 (Swanstrom, 2008, page 4). These include a gamut of disciplines such as urban and regional studies, socio-technical studies (Janssen et al., 2006), public policy theories (John, 2003), disaster studies (Vale and Campanella, 2005) and urban planning (Davoudi, 2012b; Wilkinson, 2012; Coaffee and Lee, 2016).

Much of these works are theoretically engaged with the *socio-ecological* or *evolutionary* understanding of resilience whereby resilience is not about a return to normality, but rather about 'the ability of complex socio-ecological systems to change, adapt, and crucially, transform in response to stresses and strains' (Davoudi, 2012b, drawing on Carpenter et al., 2005). Systems are defined as 'complex, non-linear, and self-organising, permeated by uncertainty and discontinuities' (Berkes and Folke, 1998, page 12). Resilience is used as a concept to unify ecological and social systems. It is advocated by the Resilience Alliance, among others, as a *general* systems theory which can integrate society, economy and ecology. This totality, dubbed *Panarchy*, is defined as: 'The structure in which systems, including those of nature (e.g. forests) and of humans (e.g. capitalism), as well as combined human-natural systems (e.g. institutions that govern natural resources use such as the Forest Service), are interlinked in continual adaptive cycles of growth, accumulation, restructuring, and renewal' (Gunderson and Holling, 2002, cover text).

The Panarchy model bundles ecological and social domains in a common conceptual and modelling framework (Walker et al., 2004). In response to some of the paradoxes of resilience thinking (such as efficiency versus redundancy, flexibility versus rigidity, learning versus uncertainty, constancy versus change), it suggests that systems function in a series of nested adaptive cycles that interact at multiple scales, speeds and timeframes. The description of the Panarchy resonates closely with the description of assemblages, as discussed above. It is argued that contingent and unknowable outcomes of complex systems require flexible forms of adaptability. It requires 'a qualitative capacity to devise systems that can absorb and accommodate future events in whatever unexpected form they may take' (Holling, 1973, page 21).

Ecologists are not alone in their desire to build general system theories. The economist Vilfredo Pareto (who introduced the principle of Pareto efficiency) also declared: 'My wish is to construct a system of sociology on the model of celestial mechanics, physics and chemistry' (Pareto, 1935, page 16). Other noteworthy sociologists who have been inspired by the natural sciences and systems theory are Talcott Parsons and Niklas Luhmann. Similarly, Fredrich Hayek, the architect of neoliberalism, drew on complexity science to develop his theory of 'spontaneous order', advocating that social order emerges from the interaction of self-serving individuals who rationally utilise the price systems to adjust their plans (Hayek, 1976). In his Nobel Prize speech Hayek declared that 'the social sciences, like much of biology but unlike most fields of the physical sciences, have to deal with structures of *essential* complexity' (1974, original emphasis).

Although the above systems-based readings of society have been heavily criticised by many contemporary social scientists, their conceptualisation of 'the social' has found renewed currency in the socio-ecological resilience literature and helped the transition of resilience from ecology to society (Olsson et al., 2015). Yet, the transition has remained conceptually problematic and normative contested. The conceptual shortcomings are evident in the debates about key features of resilience thinking including: ontology and boundary, thresholds and feedback mechanisms, functions, emergence and self-organisation (Davoudi, 2012b; Davidson, 2010; Olsson et al., 2015). The Panarchy model and its adaptive cycles are overdeterministic and do not allow for human agency, ingenuity, technology and capacity to undertake collective and goal-oriented action to break the cycles.

The normative contestation arises from the normalisation and essentialisation of resilience as a neutral and universal public good. However, no matter how it is defined – as an inherent or acquired property or as a process – resilience is not a neutral concept. It carries multiple and conflicting values and is infused with relations of power and politics. Thus, questions of resilience for whom, 'resilience from what to what, and who gets to decide?' (Porter and Davoudi, 2012, page 331) remain at the heart of the politics of resilience. For example, a study of water conflict in the German state of Brandenburg, in which Sonderhaus and Moss (2014, page 172) show how diverse constructions of vulnerability to water scarcity are translated into conflictual strategies of resilience, can be summed up as: 'your resilience is my vulnerability'. Another contested area is the advocacy of resilience as a 'buffer capacity for preserving what we have and recovering to where we were' (Folke et al., 2010, page 20). The pre-adversity order is often prescribed as the 'normal' status to which resilient individuals and societies should seek to return. This tends to foreclose debates about what is desirable for whom, who gets to decide and why not seek a new 'normal'. Julie Hernandez's (2009) account of New Orleans after Hurricane Katrina is an example of why bouncing back to where we were is not desirable if the prior condition is an unequal society with high levels of deprivation. Worst still is a regressive bounce-forth which, to a large extent, happened in New Orleans. The city transformed itself into something different from what it was socially, demographically and politically but the change has hardly been progressive (Tierney, 2015). Public housing was eliminated, public schools were replaced with charter schools, a large segment of the black population could never return to the city, no-bid contracts were allocated to big businesses and the city became awash with creating entrepreneurialism (Adams, 2012; Gotham, 2012).

The morphing of resilience from a descriptive concept in ecology to a normative one in society becomes prominent when it applied to individuals, communities and cities. While its application to individuals has a long history in the field of psychology, as mentioned earlier, its prefix to the term communities is more recent and often attributed to the work of Tobin (1999). It is worth mentioning that although the resilience literature frequently uses the term socio-ecological, the 'social' tends to be replaced by the 'community' or 'individual' in the application of the concept. As shown by Downes et al.'s (2013) quantitative, meta-analyses of 197 published papers on resilience, the predominant focus is on the individual level. One possible explanation for this scaling down is ontological, as some ecologists and social scientists define societies as 'systems' made of the sum of their subsystems such as communities and individuals. This implicitly means that resilient individuals add up to resilient communities and resilient communities add up to resilient societies. Such an understanding, however, stands against both complexity and assemblage theories which consider the whole as exceeding its constituent parts.

Another possible explanation is ideological and political (Davoudi, 2016 and 2017). The emphasis on individuals overlaps with the dominant liberal view of society and the place of the individual within it. As Norberto Bobbio (1990, page 43) put it, this 'liberal individualism' 'amputates the individual from the organic body ... plunges him into the unknown and perilous world of the struggle for survival'. Related to this atomistic view of individuals is the individualisation of responsibility which is captured in the frequently used notion of self-reliance. As a misinterpretation of self-organisation in complex systems, self-reliance advocates that individuals and communities should 'pull themselves up by their bootstraps and reinvent themselves in the face of external challenges' (Swanstrom, 2008, page 10). It claims that resilience building requires the withdrawal of welfare systems because, 'if the Government takes greater responsibility for risks in the community, it may feel under pressure to take increasingly more responsibility, thereby eroding community resilience' (Risk and Regulation Advisory Council, 2009, page 6).

Self-reliance has become the discursive bedrock of the resilience machine and the key collective/expressive component of the assemblage. Its power derives from providing a common narrative for an otherwise diverse set of actors with conflicting interests. It chimes not only with the liberal understanding of free yet responsibilised individuals, but also the conservative promotion of self-help, as well as the communitarian advocacy for self-sufficiency. Multiple contrasting values are contingently co-aligned to produce the well-oiled resilience machine capable of convincing people that self-reliance is 'a common sense, neutral and universal measure of the resilient self; one that a responsible citizen should aspire to in the face of radical uncertainties' (Davoudi, 2016, page 162). Those failing to board the resilience machine and abide by its moral advocacy are stigmatised as not being fit enough to survive in the turbulent world and its perceived unavoidable threats.

The effectiveness of the devolved and individualised responsibility is clearly manifested in the changes to disaster risk management whereby resilience has replaced vulnerability as the main discourse. Victims of disasters are cast as heroes of resiliency. The shift in discourse is mirrored in the shift in visual representation. For example, the cover of the United Nations 2004 report on resilient cities shows a single, vulnerable black woman carrying a child in her arms, while the cover of its 2010 report shows a group of resilient, smiling men who are rebuilding a community clinic. To paraphrase Logan and Molotch (1987, page 62), the resilience machine appropriates sources of civic pride, such as the ability to withstand adversities, for their own legitimation, by encouraging a sense of heroism and channelling that into activities that are consistent with responsibilisation agendas. Often, this strategy is mobilised through what Molotch (1976, pages 314–15) calls a revamped 'community "we feeling"'.

However, as assemblage and complexity theories teach us, there are always multidimensional 'possibility spaces' and 'lines of flight' within assemblages. This is reflected in the notion of 'assemblage diagram' defined as 'the structure of the possibility space associated with assemblage dispositions' (DeLanda, 2016, page 5). Hence, despite the depoliticising effects of the resilience machine and its technical and managerial language, agonistic politics remains alive and flourishing through the cracks in the concrete. A powerful sign of *resisting* is the 'Stop calling me resilient' campaign which was launched by Tracie Washington, president of the Louisiana Justice Institute, in reaction to media and policy makers' response to the victims of Hurricane Katrina and the BP oil spill. In an interview with Josh Feldman, Washington demanded:

> Stop calling me resilient, because every time you say, 'Oh, they're resilient', it actually means you can do something else, something new to my community ... We were not born to be resilient; we are conditioned to be resilient. I don't want to be resilient ... I want to fix the things that create the need for us to be resilient in the first place.
>
> *(Feldman, 2015)*

The campaign has had widespread appeal and the above words have found their way onto posters and murals in different cities including a large mural in the city of Belfast. It is a testimony to the existence of what Connolly (2005) calls 'counter machine'. It reflects the complex and dynamic politics of the resilience machine whose components are in a resonant relationship whereby 'unconnected or loosely associated elements *fold, bend, blend, emulsify, and dissolve into each other*, forging a qualitative assemblage' (Connolly, 2005, page 870, original emphasis).

Conclusion

The resilience machine is deployed at multiple, interconnected scales ranging from individual bodies to communities and cities. To understand its 'anatomy' we need to explore the imaginaries and rationalities that lubricate its technologies and materialities. We need to investigate the interconnected discourses, policies, practices and understandings of resilience that enable it to be articulated in the administration of life in human and non-human bodies, city planning, urban governance and disaster response. We understand these rationalities and technologies that govern resilience as intimately interconnected. The diverse machinic (materials) and collective (expressions) components of resilience are intertwined in a resonance relationship which enable them to forge a powerful assemblage. The ensuing resilience machine operates with such force that its effects appear natural, predetermined and incontestable. Resilience is mobilised and territorialised (institutionalised/socialised) within and through political, economic and environmental discourses and expressive acts. It works across diverse individuals, communities, cities and corporate contexts with significant human and environmental consequences.

Thinking through the anatomy of the resilience machine might be useful in considering future work that examines resilience as governmentality. By examining machine learning and the way that it operates as its own form of assemblage, infecting nearly every element of modern life, it is also possible to explore the progressive and liberating potentials of resilience. The evolutionary and contingent nature of the resilience machine should be understood as a source of both limitations and opportunities. Regressively, if the machine is perpetuated as a tool of economic expansion and capitalist accumulation, the most vulnerable people, places and environments will continue to suffer. Alternatively, the adaptability of the machine keeps multiple 'possibility spaces' for a governmentality that embraces the ideals of justice, equity and freedom from oppression.

References

Adams, V. (2012). 'The other road to serfdom: recovery by the market and the affect economy in New Orleans'. *Public Culture*, 24: 185–216.
Alexander, D.E. (2013). 'Resilience and disaster risk reduction: an etymological journey'. *Natural Hazards Earth System Science*, 13: 2707–2716.

Berkes, F. and Folke, C. (1998). *Linking social and ecological systems: management practices and social mechanisms for building resilience*. Cambridge: Cambridge University Press.

Bloch, D.A., Silber, E. and Perry, S.E. (1956). 'Some factors in the emotional reaction of children to disaster'. *American Journal of Psychiatry*, 113: 416–422.

Boardman, P. (1978). *Worlds of Patrick Geddes: biologist, town planner, re-educator, peace warrior*. London: Routledge.

Bobbio, N. (1990). *Liberalism and democracy*. London: Verso.

Capra, F. (1982). *The turning point: science, society, and the rising culture*. New York: Bantam Books.

Carpenter, S.R., Westley, F. and Turner, G. (2005). 'Surrogates for resilience of social-ecological systems'. *Ecosystems*, 8(8): 941–944.

Carrington, D. (2016). 'Climate change will stir unimaginable refugee crisis, says military'. *Guardian*, 1 December. Available at: www.theguardian.com/environment/2016/dec/01/climate-change-trigger-unimaginable-refugee-crisis-senior-military

Clifford, T. (1975). *The political machine: an American institution*. New York: Vantage Press.

Coaffee, J. and Lee, P. (2016). *Urban resilience: planning for risk crisis and uncertainty*. London: Palgrave Macmillan.

Connolly, W. (2005). 'The evangelical-capitalist resonance machine'. *Political Theory*, 33(6): 869–886.

Davidson, D.J. (2010). 'The applicability of the concept of resilience to social systems: some sources of optimism and nagging doubts'. *Society and Natural Resources*, 23: 1135–1149.

Davoudi, S. (2012a). 'Climate risk and security: new meanings of "the environment" in the English planning system'. *European Planning Studies*, 20(1): 49–69.

Davoudi, S. (2012b). 'Resilience: a bridging concept or a dead end?'. *Planning Theory and Practice*, 13(2): 299–307.

Davoudi, S. (2016). 'Resilience and governmentality of unknowns', in M. Bevir (ed.), *Governmentality after neoliberalism*. New York: Routledge, pp. 152–171.

Davoudi, S. (2017). 'Self-reliant resiliency and neoliberal mentality: a critical reflection', in E. Trell, B. Restemeyer, M. Bakema and B. van Hoven (eds), *Governing for resilience in vulnerable places*. London: Routledge, pp. 1–7.

DeLanda, M. (2016). *Assemblage Theory*. Edinburgh: Edinburgh University Press.

Deleuze, G. and Guattari, F. (1987). *A thousand plateaus*. Minneapolis, MN: University of Minneapolis Press.

Deleuze, G. and Parnet, C. (2002). *Dialogues II*. New York: Columbia University Press.

Dittmer, J. (2014). 'Geopolitical assemblages and complexity'. *Progress in Human Geography*, 38(3): 385–401.

Downes, B.J., Miller, F., Barnett, J., Glaister, A. and Ellemor, H. (2013). 'How do we know about resilience? An analysis of empirical research on resilience, and implications for interdisciplinary praxis'. *Environmental Research Letter*. Available at: stacks.iop.org/ERL/8/014041

Feldman, J. (2015). 'MSNBC guest: stop using the word "resilient" to describe Katrina victims'. *Mediaite*, 29 August. Available at: www.mediaite.com/tv/msnbc-guest-stop-using-the-word-resilient-to-describe-katrina-victims/

Folke, C., Carpenter, S., Walker, B., Scheffer, M., Chapin, T. and Rockstrom, J. (2010). 'Resilience thinking: integrating resilience, adaptability and transformability'. *Ecology and Society*, 15(4): 20–28.

Garmezy, N., Masten, A.S. and Tellegen, A. (1984). 'The study of stress and competence in children: a building block for developmental psychopathology'. *Child Development*, 55: 97–111.

Gotham, K. (2012). 'Disaster, inc.: privatization and the post-Katrina rebuilding in New Orleans'. *Perspectives on Politics*, 10: 633–646.
Gunderson, L.H. and Holling, C.S. (eds) (2002). *Panarchy: understanding transformations in human and natural systems*. Washington, DC: Island Press.
Hayek, F.A. (1974). 'The pretence of knowledge'. Acceptance speech upon the award of the Sverige Rigsbank Prize in Economics in Memory of Alfred Nobel, Salzburg, 11 December.
Hayek, F.A. (1976). *Law, legislation and liberty, the mirage of social justice* (Vol. 2). London: Routledge and Kegan Paul.
Hernandez, J. (2009). 'The long way home: New Orleans, still a disaster area three years after Hurricane Katrina'. *L'Espace géographique*, 2(38): 124–138.
Holling, C.S. (1973). 'Resilience and stability of ecological systems'. *Annual Review of Ecological Systems*, 4: 1–23.
Holling, C.S. (1996). 'Engineering resilience versus ecological resilience', in P.C. Schulze (ed.), *Engineering within ecological constraints*. Washington, DC: National Academy Press, pp. 31–45.
Hornborg, A. (2006). *The power of the machine: global inequalities of economy, technology, and environment*. Walnut Creek, CA: AltaMira Press.
Janssen, M., Schoon, M., Ke, W. and Borner, K. (2006). 'Scholarly networks on resilience, vulnerability and adaptation within the human dimensions of global environmental change'. *Global Environmental Change*, 16(3): 240–252.
John, P. (2003). 'Is there life after policy streams, advocacy coalition and punctuations: using evolutionary theory to explain policy change?'. *Policy Studies Journal*, 31(4): 481–498.
Lasswell, H. (1936). *Politics: who gets what, when, how?* New York: McGraw-Hill.
Latour, B. (2005). *Reassembling the social: an introduction to actor-network theory*. Oxford: Oxford University Press.
Lauermann, J. (2016). 'Municipal statecraft: revisiting the geographies of the entrepreneurial city'. *Progress in Human Geography* 42(2): 205–224.
Logan, J.R. and Molotch, H. (1987). *Urban fortunes: the political economy of place*. Oakland, CA: University of California Press.
Malins, P. (2004). 'Machinic assemblages: Deleuze, Guattari and an ethico-aesthetics of drug use'. *Janus Head*, 7(1): 84–104.
Manson, S. (2001). 'Simplifying complexity: a review of complexity theory'. *Geoforum*, 32: 405–414.
McCann, E. (2013). 'Policy boosterism, policy mobilities, and the extrospective city'. *Urban Geography*, 34: 5–29.
McFarlane, C. (2011). 'Assemblage and critical urbanism'. *City*, 15(2): 204–224.
Molotch, H. (1976). 'The city as a growth machine: toward a political economy of place'. *American Journal of Sociology*, 82: 309–332.
Molotch, H. (1999). 'Growth machine links: up, down, and across', in A.E.G. Jonas and D. Wilson (eds), *The urban growth machine, critical perspectives two decades later*. New York: State University of New York Press, 247–267.
Olsson, L., Jerneck, A., Thoren, H., Persson, J. and O'Byrne, D. (2015). 'Why resilience is unappealing to social science: theoretical and empirical investigations of the scientific use of resilience'. *Science Advances*, 1(e1400217): 1–20.
Oxfam International (2017). 'Just 8 men own same wealth as half the world'. 16 January. Available at: www.oxfam.org/en/pressroom/pressreleases/2017-01-16/just-8-men-own-same-wealth-half-world

Oxford Dictionary (2017). Available at: https://en.oxforddictionaries.com/definition/machine
Pareto, V. (1935). *The mind and society*. New York: Harcourt, Brace and Company.
Peck, J. (2016). Presentation at the Regional Studies Association's winter conference, New pressures on cities and regions, 24–25 November, London.
Porter, L. and Davoudi, S. (2012). 'The politics of resilience for planning: a cautionary note'. *Planning Theory and Practice*, 13(2): 329–333.
Risk and Regulation Advisory Council (2009). *Building resilient communities, from ideas to sustainable action*. London: RRAC.
Roth, K. (2017). *Human Rights Watch. World Report 2017*. Available at: www.hrw.org/world-report/2017/country-chapters/dangerous-rise-of-populism
Schindler, S. (2016). 'Detroit after bankruptcy: a case of degrowth machine politics'. *Urban Studies*, 53: 818–836.
Smith, M.P., Koikkalainen, S. and CasanuevaL.J. (2014). 'The oligarchic diffusion of public policy: deploying the Mexican "magic bullet" to combat poverty in New York City'. *Urban Affairs Review*, 50: 3–33.
Sonderhaus, F. and Moss, T. (2014). 'Your resilience is my vulnerability: "rules in use" in a local water conflict'. *Social Sciences*, 3: 172–192.
Surborg, B., Van Wynsberghe, R. and Wyly, E. (2008). 'Mapping the Olympic growth machine'. *City*, 12: 341–355.
Swanstrom, T. (2008). *Regional resilience: a critical examination of the ecological framework*. Berkeley, CA: Institute of Urban and Regional Development, University of California.
Tampio, N. (2009). 'Assemblages and the multitude: Deleuze, Hardt, Negri, and the postmodern left'. *European Journal of Political Theory*, 8: 383–400.
Taylor, C. (2004). *Modern social imaginaries*. Durham, NC: Duke University Press.
Thrift, N. (1999). 'The place of complexity'. *Theory, Culture and Society*, 16: 31–69.
Tierney, K. (2015). 'Resilience and the neoliberal project: discourses and practices – and Katrina'. *American Behavioural Scientists*, 59: 1327–1342.
Tobin, G.A. (1999). 'Sustainability and community resilience: the holy grail of hazards planning?' *Global Environmental Change Part B: Environmental Hazards*, 1: 13–25.
Vale, L.J. and Campanella, T.J. (eds) (2005). *The resilient city: how modern cities recover from disaster*. New York: Oxford University Press.
Walker, B., Holling, C.S., Carpenter, S. and Kinzig, A. (2004). 'Resilience, adaptability and transformability in social-ecological systems'. *Ecology and Society*, 9(2): 5. Available at: www.ecologyandsociety.org/vol9/iss2/art5/
Wilkinson, C. (2012). 'Urban resilience: what does it mean in planning practice?' *Planning Theory and Practice*, 13(2): 319–324.
World Economic Forum (2017). *Insight report: the global risks report 2017* (12th edn). Available at: www3.weforum.org/docs/GRR17_Report_web.pdf
Zolli, A. (2012). 'Learning to bounce back'. *New York Times*, 2 November. Available at: www.nytimes.com/2012/11/03/opinion/forget-sustainability-its-about-resilience.html

2

SECURING THE IMAGINATION
The Politics of the Resilient Self

Julian Reid

As a culture we are saturated today by discourses around the need to develop the self. In fact discourses around selfhood can be seen to be driving wider social and political discourses around development on a more or less global scale. The discourse on 'the resilient self' is a case in point. Leading psychologists of resilience claim responsibility not just for developing the concept from its psychological origins into the international political and social framework it has now become, but for the peace and reconciliation in former war-torn countries where resilience is now said to exist. Within resilience, however, lurks another property and capacity of the self, that of imagination. Imagination is said by psychologists to play a crucial role in the recovery, for example, of human beings from traumatic experiences and their development of resilience. For abused children, especially, recovery and development are said to require the work of imagination, as hurt creates the images of a better future and the pleasure of such images becomes linked with painful realities, enabling them to withstand the present. It is even possible, some psychologists of resilience maintain, that the torment of the present heightens the need to imagine a future and thus increases the very powers and potentials of imagination itself. How, then, to theorise and understand the work which imagination performs on social and political scales? In this chapter I want to consider, critique and extend such psychological accounts of the function of imagination for the purpose of developing a better understanding of the politics of resilience. Images are, I argue, while untrue and in a certain sense inferior to the real, nevertheless things which human beings need in order to be able to act collectively upon the real, and to change the very nature of their political and social circumstances. But while resilience provides scope for the function of imagination in enabling human beings to survive, it is nevertheless, as a discourse, also based upon a highly circumscribed imaginary, the limits of which are defined by survivability as such. Imagination can either contribute to the survival strategies with which human beings attempt to care for themselves in the face of ordeals and traumas, or it can, more ambitiously, seek to create an image of the self, existing free from the possibility and necessity of a life of endless trauma and struggle. It is this latter task that I will argue deserves the greater exploration today.

The Image of the Resilient Self

The image of the self as that which must struggle to survive, that which is in need of care, and always in danger, of dispossession and harm, is a fairly universal one today. One encounters it not just in the popular self-help manuals on resilience that clutter the shelves of bookstores but in fairly philosophical works on the self such as Judith Butler and Athena Athanasiou's recent book, *Dispossession: the performative in the political*. As they argue in a manner that expresses the fundamental principle of the governing orthodoxy on the self:

> We cannot understand ourselves without in some ways giving up on the notion that the self is the ground and the cause of its own experience ... dispossession establishes the self as social, as passionate, that is, as driven by passions it cannot fully consciously ground or know, as dependent on environments and others who sustain and even motivate the life of the self itself.
>
> *(Butler and Athanasiou, 2013, page 4)*

Being dependent on others who may alternately sustain or deprive us means, Butler and Athanasiou argue, that others always 'hold a certain power over our very survival' (Butler and Athanasiou, 2013, page 4). The self is thus to be considered as vulnerable and only ever able to struggle to survive in its dependence on others. How it survives, how it manifests resilience, in the context of its vulnerability to and dependence on others that may either sustain or deprive it of life is a question of what Butler and Athanasiou call self-poetics; the arts deployed to make and produce the self by subjects in struggle with others who hold power over their survival. At work in this theory of the self, then, is a highly particular image of the self. An image of the self as vulnerable, as a subject of constant and necessary struggle, one of which resilience is demanded, in the context of the claim that it must consistently remake itself to survive.

One of the major criticisms to be made of resilience is that, while it provides scope for the function of imagination in conceiving how human beings survive, it also bases itself upon a highly circumscribed imaginary the limits of which are defined by survivability as such. The image of the human governing this imaginary is that of the survivor itself. Such that life cannot be imagined as anything other than a game of mere survival. The discourse of resilience, whether in its popular self-help form or the philosophical form in which we encounter it, in the work of Judith Butler, credits human beings with imagination, but always and only in the interests of survival. Images are conceived as forms of technology, instruments for getting humans out of situations of struggle and servitude, but only so that they might live to fight another day. The function of imagination in this discourse is one that serves to embellish the image of the human as survivor.

There is no better example of this than the work of the French psychologist Boris Cyrulnik, whose book *Resilience: how your inner strength can set you free from the past* is sold on the claim that 'his work has been credited with helping France heal the wounds left by the Second World War' (Cyrulnik, 2009). For Cyrulnik, imagination plays a crucial role in the recovery of human beings from traumatic experiences, and for the 'knitting together' of a feeling of selfhood, which he argues is the major factor in determining aptitudes for resilience (page 19). Cyrulnik describes a demoralised child, who having suffered abuse, experiences confusion and hurt mingled with hope. 'As soon as he is hurt, the child begins to dream of a better future … and because the pleasure of the dream becomes linked with the painful reality, he is able to withstand it' (page 4). It is even possible, Cyrulnik maintains, 'that the torment heightens the need to imagine a future' (page 4). Resilience, he argues, is a word which can help us understand the mystery of survival; the way that is by which human beings, amid great suffering, are able to create an image of a future, where free from the pains and oppressions of the present, life will be lived differently, and of course, the function of that image in leading the way, providing a light, by which the human beings in question find their way to a future which enables them to survive the present.

There is a further way in which the capacity for resilience is said by Cyrulnik to enlarge the imagination. For the subject who is capable of imagining the possibility of disasters on the horizon is also the subject more capable of imagining events that exceed the norm. As Cyrulnik argues, we live in a culture where we are taught to believe that the best of all possible worlds is one in which everything turns out best. Against what he argues to be a dominant cultural norm, Cyrulnik opposes, precisely, the figure of the survivor. He or she who knows, from experience, that things don't always turn out well, and that preparedness for extreme suffering is necessary. 'Don't fall for it,' the survivor warns, 'the next disaster is on its way.' This unfounded optimism as to the image of the future, which supposedly defines our culture, expresses the limits of the imaginations of those who have shaped it. It is to them that Cyrulnik's survivor speaks; 'you cannot imagine anything that is out of the ordinary. You see anything that deviates from the norm as an act of aggression' (Cyrulnik, 2009, pages 25–6).

Is it the case, as Cyrulnik maintains, that our culture is geared to preventing us from imagining that which is out of the ordinary, so beyond the norm, including potential disasters? The *9/11 Commission Report* (2004), prepared at the request of the then United States president, George W. Bush, in response to the surprise attacks made upon the United States in 2001 by Al-Qaeda, berated the problem that 'imagination is not a gift usually associated with bureaucracies' (page 344). The military history of the United States, it noted, is defined by recurrent failures of imagination, most notably the failure to have imagined the Japanese bombing of Pearl Harbor; an event that in spite of its calculable potential was also not reckoned upon (9/11 Commission Report, 2004). 'It is therefore crucial,' the report concluded in the wake of the apparent exposure of the lack in imagination of the United States military, given its failure to have anticipated the 9/11 attack, 'to find a way of routinizing, even bureaucratizing, the exercise of imagination' (page 334). In confronting the threat of terror, here in the 21st century, the United States security and strategic community considers itself faced with an enemy possessive of greater capacities of and for imagination than itself. Nobody epitomised the intensity of that threat, the *9/11 Commission Report* reported, more than Khalid Sheikh Mohammed himself, the principal architect of the 9/11 attacks. Alongside his technical aptitudes and managerial skills, Khalid Sheikh 'applied his imagination' to constructing an 'extraordinary array of terrorist schemes' the report concluded (page 145). The image the report constructs is not that of a culture that condemns the subject who imagines too excessively, but one which, like Cyrulnik's survivor, condemns it for not imagining enough. Likewise, it constructs the image of a culture that fears another, more adept at applying imagination than itself, and on which account, it demands more imagination, greater imagination, in order to overhaul the threat posed by enemies' possessive of such rampant imagination.[1] This is precisely what the routinisation and bureaucratisation of imagination it calls for is about.

The interest of power and government in accessing the imaginations of its subjects is, of course, not new. Back in the 17th century, the great English philosopher Thomas Hobbes lamented the ways in which crafty governments 'abuse the simple people' by manipulating their faculties of imagination (Hobbes, 1993, page 93). For Hobbes, political problems stem from human beings possessing too much imagination, and being too willing to invest credibility in this or that absurd image, leading to belief in false prophesies and generating superstitious fears. Modern theories of state power have proceeded in his wake on the assumption that the imaginations of modern subjects are vulnerable to such manipulation. The dangers and threats which states construct do their work upon the imaginations of societies, building the fears on which the power of states rests and grows. Critical agency, it has long been presumed, depends on challenging these strategies of construction, releasing imagination from its manipulation by power, and exposing the imaginary nature of the threats and dangers which state power creates, by rationalising the world better.[2] What can and should critical theorists say, then, about this present juncture in which the image of danger by which the state constructs its threats is that of an enemy itself possessive of extraordinary imagination? What also can and ought critical theorists say about this present in which state powers lament the lack of imagination of their own citizens, bureaucracies and cultures?

For political theorist Mark Neocleous this concern of the United States security and strategic communities to secure imagination is a frightening development. For it indicates the will of the state to 'police' and even 'colonise' our political imaginations through the development of new state strategies of resilience.[3] It is, he argues, a strategy that embodies the age-old Hobbesian premise of state power, fulfilling the need of the state to manipulate the imaginations and fears of its subjects (Neocleous, 2016, page 2). Indeed for Neocleous, as for many other critical theorists of the 'War on Terror', we are living through an era in which states possess remarkable powers of manipulation and construction when it comes to the question of danger and threat, and the societies of such states are powerless and docile in the face of state strategies.

I do not follow Neocleous in his assumption that this expresses a crude will of 'the state' to capture our political imaginations. How then should we understand the wider discourses on resilience and the ways in which imagination is said to be a resource for resilience by psychologists and other forms of theorists and practitioners? Have they simply been brainwashed by 'the state'? Are they its ideological proxies? Instead of resorting to such crude formulations we have to address the way in which the development of the concept of resilience, across and throughout not just the social sciences, but also the humanities and life sciences, as well as all manner of other forms of thinking about resilience entails this revalorisation of imagination. The incorporation of a concern for promoting imagination in American security doctrine is just a part of this revalorisation.

Critical thinkers, not belonging to the US state department or working within the Pentagon, but of the political Left, concerned with the limits of the imaginaries through which the species is attempting to survive and proffer resilience in the face of potentially catastrophic climate change, for example, also insist on the fundamentality of imagination for our abilities to manifest resilience and adapt to the new realities. As Kathryn Yusoff and Jennifer Gabrys argue, resilience 'may not be best realized – or imagined – through depoliticized capsules for survival, but rather through more thorough-going encounters with the social and political connections that make survival and adaptation possible – and ethical. Such encounters inevitably raise questions about the scale at which we imagine adaptation to be viable' (Yusoff and Gabrys, 2011, pages 516–34). The destructive nature of man-made climate change raises the question, they go on to argue, of 'how we imagine and understand the collective human condition, the longevity and sustainability of homo sapiens' (page 529).

'The work of the imagination' challenges us to extend our conceptions of what the human is and 'perhaps the greatest work of the imagination' is its power to 'imagine how we might be otherwise' in the context of destructions the species has wrought upon the planet it depends on to survive and, as a consequence, on other life forms (Yusoff and Gabrys, 2011). 'Our ability to imagine other possibilities, to embrace decidedly different futures with creativity and resolve, to learn to let go of the sense of permanence we may have felt about certain landscapes that have seemed to be always so, and to embrace change, is paramount,' they conclude, 'to building resilience and adaptive capacity' (page 529). 'Requisite imagination,' as other prominent proponents of resilience have argued, the ability 'to expect the unexpected, and to look for more than just the obvious … is a *sine qua non* of resilience' (Hollnagel et al., 2006, page 349).

This is just one example of how diffuse thinking about relations between resilience and imagination is today. There is a widely held belief, expressed from different positions spanning the political spectrum, articulated by use of a wildly different range of discourses, drawing on different forms of scientific knowledge, addressing different social and political problems, as to the need to galvanise imagination, in order to build resilience. Beneath the regime of resilience which the author (and others) have analysed the origins and directions of, there is a growing regime of imagination (for the analysis of resilience see especially, Evans and Reid, 2013; Chandler, 2013).

The function of this regime of imagination is one of limiting the political imaginations of whoever is subject to it. In this sense it enjoys deep continuities with the strategy of state power which Hobbes identified back in the 17th century, and which continues to be of importance for our understanding of how power works through the manipulation and mobilisation of the human imagination. However, this is not a strategy grounded simply in a capacity to conjure up fake or misleading images of threat. Nor does it emanate simplistically from 'the state'. It does not just exploit the overzealous attributes of imagination that can lead human beings to imagine danger where it is not. Instead, it functions by circumscribing our imaginations, limiting its motions within the boundaries of the regime of resilience. The problem for critique, therefore, is not one of pointing out how and where imaginary threats masquerade as real. It is one of mapping the imaginary in and through which imagination is captured. How, consequently, might we liberate imagination from its subjugation within this dominating imaginary such that it can create images which do political work of another kind? Where, even, can we find and draw succour from other imaginaries of life? This is the problem and question motivating the inquiry of this chapter.

The Limits of Imagination

The theorisation of imagination, and its functions in human experience, has been absolutely central to Western philosophy, from its inceptions in the classical eras of ancient Greece and Rome. In one dominant tradition, extending from Plato, images, the products of imagination, have tended to be condemned as inferior to the real, and sources of deception; barriers to the discovery of truth. But in other more minor traditions they have been understood as things human beings need in order to be able to act collectively upon the real, and to change the very nature of their individual, political and social circumstances. But grappling with the politics of imagination, particularly the roles it plays today in these new discourses of resilience, requires us to delineate the differences and relationships between imagination, images and the imaginaries in which they function. For while resilience provides scope for the function of imagination in enabling human beings to survive, it is nevertheless, as a discourse, also based upon a highly circumscribed imaginary, the limits of which are defined by survivability as such.

Imagination is limited if the images it produces only serve to contribute to the strengths of a dominant or governing imaginary. There is nothing radical or progressive or alternative in insisting on the power of imagination as such. Imagination is something everybody has. Nobody, we can suppose, lives without producing images. The critical question is what are its limits? What are the boundaries which determine the imaginary in which a given imagination functions? Selfhood, as contemporary regimes of resilience demonstrate aptly, requires the making of images of the self to work. Likewise, challenging any given regime of the self requires an act of imagination that transcends the imaginary in which that self is captured. In this sense imagination can either contribute to the survival strategies with which human beings attempt to care for themselves in the face of ordeals and traumas, or it can, more ambitiously seek to create an image of the self, existing free from the possibility and necessity of a life of endless trauma and struggle. It is this task of imagining in ways that will challenge the imaginary in which the resilient self is presently captured which I believe to be the critical task.

Not only is the function of imagination in servicing the survival strategies of subjects, be they states, societies or individuals, circumscribed by an imaginary of survival as such. It is questionable whether such functions of imagination, in servicing survival, qualify imagination as fully human. Even animals utilise their imaginations to survive. The Canadian philosopher Adam Morton, for example, argues that it is a capacity of mice; 'a fearful mouse imagines the dangers facing her, and people can imagine in ways that need little refined human capacity' (Morton, 2013, pages 3–4). His argument runs that the exercise of imagination does not even require images, for 'images and words are just one way in which we can grasp possibilities that might be important'. When we imagine, he argues, we simply 'represent something to ourselves; a fact, a thing, or a possibility' (Morton, 2013, page 8). Imagination thus conceived is only 'a process of searching for representations suitable for a specific purpose' (page 9). It is always ends-oriented as 'all imagining has a purpose' (page 10).

I don't accept Morton's argument in its entirety. Such a way of conceiving imagination leaves no room for discrimination between imagination and the making of representations as such. Indeed, it begs multiple questions concerning the differences between imagination and representation, questions that go unaddressed in Morton's work. Such a way of conceptualising imagination effectively debases all that is specifically human to it. Imagination is nothing if it does not produce an image, and an image is never simply a representation. If it were simply the case that imagination was the faculty for the making of representations, and representations that are always end-specific, then we would have no need for the concept of imagination, or indeed of the image. We could speak simply of representation. A mouse no doubt does make 'representations of possible situations' to itself when it 'anticipates carnivores leaping from unexpected places' (Morton, 2013, page 13) and responds with a dash for shelter, but such representations are unlikely to ever exceed its mousiness. Were it to represent itself to itself, as a carnivore, of equivalent ferocity and guile, to that which it perceives lurking, possibly, around the corner, then it would be engaged in an exercise of imagination, and not simply representation. There is, in other words, an excess to the image which differentiates it from a mere representation. I am happy to concede the concept of representation to the games we humans also play out, throughout life, which help us in 'setting in place a structure of responses to possible threats' (page 13). Such embodied and mental practices are integral to the adaptive life of every living being, human and otherwise. But imagination, I believe, is quite different. An image, fully conceived, is more likely to interfere with and corrupt whatever means–ends calculations we are engaged with, as much as our imaginations are in conflict with our faculties for reasoned belief, perception and action. The imagination, 'the ruling and divine power', is as Francoise Delsarte declared, 'never governed' (Delsarte, 2011). It is what makes human beings so interesting. Imagination is never a means by which we simply adapt and respond to our environments, in some strategic way, of accommodating ourselves to whatever we perceive to be threatening. Our images have a life that is their own.

The question, then, is what is in an image? What makes the difference between images that function to enable response and adaptation to take place and images that challenge and transform? In 2010 the American director, Jim Jarmusch, released a film, *The limits of control*, to wide critical acclaim. For a full account of the film see Rice (2012). The film depicts the journey through Spain of a Mysterious Loner set upon penetrating the heavily fortified headquarters of American Empire, the history of which the film narrates in its genealogical intertwinements with that of the Spanish and other European imperial formations. Journeying deep into the Spanish countryside, Mysterious Loner breaks into the office of its president-like figure, the so-called Big Boss. Unlike most comparable thrillers the film contains no visual sequence in which the feat of the breaking into the imperial headquarters and office of its Big Boss is depicted. The viewer simply is not shown and does not know how that seemingly miraculous feat was achieved. Instead, the film cuts to the scene of confrontation between the Big Boss (played by Bill Murray) and Mysterious Loner (played by Isaach De Bankolé). The Big Boss is sitting behind his desk when he looks up and catches glance of the Loner.

BOSS: 'How the fuck did you get in here?'
LONER: 'I used my imagination.'

The scene culminates in Mysterious Loner's murder of the Big Boss, as he strangles him with the cord of his own telephone. The film has been interpreted largely as an anti-colonial and anti-American fantasy, combining as it does the motifs of contemporary American imperialism with those of historical Spanish colonialism (Rice, 2012). The overwrought power, racism and arrogance of the figure of the American is contrasted with the humility, cosmopolitanism and asceticism of the tai chi-practising Mysterious Loner. And the act of imagination by which Mysterious Loner penetrates the seemingly impenetrable fortress of the headquarters of the empire and topples its sovereign contrasts with the dependence of the Big Boss on his crude access to information.

One might say that Jarmusch's film celebrates the irreducibility of imagination, its essentially ungovernable nature, and hostility as well as superiority to political power. Indeed, this is how Jarmusch's work, and not just *The limits of control*, has been interpreted – as an homage, politically and culturally, to the power of the human imagination to transcend boundaries (Rice, 2012). But does *Limits of control* not presuppose an image of power itself, as lacking in imagination, and dependent on crude mechanisms of control, in ways that no longer describe the imaginal properties of regimes of power in a post-9/11 world? And can we be sure that Jarmusch's character, Mysterious Loner, performs his imagination in ways that go beyond its theorisation as a function in the search for representations suitable for specific purposes in the manner that Adam Morton reduces imagination to? *Limits of control* is ultimately, perhaps, a revenge movie. As Morton himself notes, a lot of emotion-driven thinking, including the kind of thinking motivated by the desire for revenge, is imaginatively creative, and in line with what is believed to be the biological function of emotions – 'to make us search for actions which will address problems of particular kinds' (Morton, 2013, pages 22–3). This involves the creation of images that enable us to act in and on the real.

Contemporary neuroscience has claimed significant advances, through the development and application of new technologies for brain imaging, in the understanding of how imagination helps us to act in and on the real. Its major lesson is that images do indeed have a function for human beings in enabling them to act more proficiently in the real. We know, for example, that when someone imagines she is dancing, the image she creates activates the same area of the brain as is activated when she actually dances in reality. The measurement of brain activity, enabled by the latest scanning devices, tells us so. Brain images of the effects of physical and imaginary movements on the brain reveal the precise areas of the brain where these two different forms of movement coincide (Frith, 2007, pages 12–13). The image created by a purely imaginal movement is of use to the human only in so far as it enables him or her to approximate towards something that is real in the world. Tests show that human beings asked to train for real events entailing real movements by first imagining them over a period of time are better able to carry them out on account of the imaginary preparation. We learn how to do things by simply imagining the movement entailed (page 106). Thus, we learn the value of the imagined movement by transforming it into a real movement.

Perhaps what neuroscience tells us, then, is that the act by which Mysterious Loner penetrated the headquarters of the American Empire and toppled Big Boss was not so miraculous at all. And perhaps, as it also indicates, the use of imagination as method for the performance of such a task, as claimed by Mysterious Loner, is not so far-fetched either. Mysterious Loner, we might suppose, learned how to perform that act by carrying out the necessary imaginary preparation. And he learned the value of the imaginary movement of the penetration of the Empire by transforming it into a real movement, rather than leaving it in the realm of fantasy. The neuroscience converges, in this sense, with the aesthetic representation of the power and indeed functional utility of imagination, in ways that make the homage to imagination at work in Jarmusch's film less romantic and transcendent than it might otherwise seem. Indeed, there are many ways in which one might compare Jarmusch's aesthetic representation of Mysterious Loner, a character equipped with imagination and dedicated to destroying American Empire, with the United States representation of Khalid Sheikh Mohammed, a man also said to have been abundant in imaginative schemes and dedicated to destroying American Empire.

The aesthetic and neuroscientific arguments for and representations of the functional utility of imagination are also verified by a third source: capitalism. It is a fact that the power of imagination and human capacity to make mental images has become integrated within the machinery of capitalism and neoliberal economies in recent years. Shoshana Zuboff's classic, *In the age of the smart machine*, written and published in the 1980s, described how the introduction of computing and electronic data changed not just the kind of work which industrial labourers were required to do, but the 'quality of mind' demanded by such work (1988, page 85). Suddenly the ability to have 'inner vision, constructed out of a combination of memory and imagination' became a key capacity demanded of the industrial labourer. The complexity and especially the abstract nature of electronic data meant that operators had to become adept at 'mental imaging' in order to be able to respond efficiently to data (page 86).

The image, we can conclude, then, from these three sources of verification (the neuroscientific, the aesthetic, the economic) is useful. Images and imagination help us to function more proficiently in the real world. In this sense they are subordinate to the real. Their value, however creative they may be, is logistical, in that they enable us to act more proficiently. Whether we are a dancer learning to dance, a mysterious loner looking to penetrate the headquarters of American Empire, or an industrial worker in a factory, they enable us to get things done. In each instance the self extracts from the image forms and forces that serve to render it useful for a specific problem-solving purpose.

It is not the image but the real, in each of these contexts, that is conceptualised and represented as the truly creative force, continually outstripping our abilities to control it, know it and even see it. Images are simply the tools with which we manage our survival in subordination to the creative forces of reality. Brain imaging studies indicate that there is a 'face area' of the brain where neural activity occurs regardless of whether the face in question is real and encountered or imagined (Frith, 2007, page 137). The face area in the brain becomes active whenever we see a face or imagine a face. In both instances, the brain creates a face. What, then, is the difference? For neuroscientists such as Chris Frith, the difference is that in encounters with real faces, the face imaged by the brain never perfectly conforms with that which is seen and received by our senses, and thus we must continually revise its image on account of the ways in which the face seen escapes its imaginary model. The imagined face on the other hand does not err. It appears to us, but requires no revision. It is perfect in itself. For Frith this indicates the 'utterly uncreative' nature of imagination. For it has 'no errors to resolve' (page 137).

As philosopher Jan Slaby has described, neuroscience has sold itself on the claim to be able to unveil the fundamental functioning principles of the brain and the central nervous system. The brain is the zone of intersection where matter supposedly meets mind, and in its unveiling of these principles, neuroscience supposedly unveils the fundaments of subjectivity, consciousness, agency – the very core of the human (Slaby, 2015, pages 16–22). Slaby identifies neuroscience and the social sciences which are influenced by it with what he calls 'a bioperfection imperative', which consists of a reframing of both life and the mind as a resource, potential and source of risk in the 'drive towards technological perfection of the human condition' (page 17). As such it is as discursive every bit as much as humanist scholarship is technically sophisticated and materially grounded. Neither the natural sciences nor the social sciences or the humanities can claim any greater or deeper grip on the real. The lesson Slaby extracts from this is the need, firstly, for a thorough repoliticisation of neuroscientific truths – one which will highlight the struggles for discursive, institutional and economic dominance within neuroscience itself, and secondly, for political interventions within these fields.

It is on this basis, building on Slaby's critique and call, that I argue we need to challenge and reverse neuroscientific formulations of the relation of the imagined to the real; this notion of the functionality of imagination, of the reduction of the image to resource in a life of endless survival, and ultimately of the subordination of the image to the real. And I believe there are good humanistic grounds on which to do so. Images are of many kinds. In effect there is no such thing as 'the image' or 'the imagination' in the ways that neuroscience and its ideologues, so powerful today in the social sciences and in the framing of governmental policies, suppose. Instead we need a typology of the many different kinds of images that exist, and the many different types of movement of which imagination is capable. To begin with it is necessary at a minimum to state the fact that images produced by the brain in response to real faces are always copy-images, reflecting an encounter and response to the real, while imagined faces are pure creations, less diluted by a relation to some other thing they have learned from in order to correct their errors. Likewise, the image produced by an imaginal movement within the brain has characteristics of its own which will be lost in the translation to real movement which occurs when real movement is the task at hand. In both instances the real functions to extract from the image forms and forces that serve to limit its range of expression. Why should this be so? We must consider the ability to reverse this conceptualisation of the image to the real, and of the profoundly human power to subordinate the real to the image, such that it is made to conform to what we imagine.

The French philosopher Gilles Deleuze created a typology of images within his work that involved discerning the specificities of many different kinds of image, including what he described as the sensory-motor image and the optical-sound image (Deleuze, 1989, pages 44–7). A sensory-motor image is one which gives rise to movement. In service of the operation of a motorised movement it extracts from reality only that which is useful for that purpose. An optical-sound image, in contrast, affects us in ways that do not lead to motorised movement. In making an optical-sound image we extract from the real, features more likely to disable our motorised capacities for action, slowing us down, sending us in directions we had not perceived possible, making us less efficient, but more attuned to layers of the real which a sensory-motor image would screen out.

Deleuze (1989) detailed the relative richness versus poverty, depth versus superficiality, of sensory-motor and optical-sound images, in ways that inverted the neuroscientific formulation of the value of images. From a strictly superficial point of view, the sensory-motor image would seem richer, in so far as it gives us access to the thing itself of which the image derives; access at least to 'the thing as it extends into the movements by which we make use of it' (page 44). In contrast the optical-sound image would seem poorer and lacking depth in so far as it involves only a selection of features of the thing itself, a 'description', which does not help us to make use of it. But the richness of the sensory-motor image is itself superficial in so far as it gives us access only to what that thing has in common with every other thing of which we make like use. It is not this particular grass that interests the herbivore, but grass in general, when it creates for itself a sensory-motor image of grass, in the event of grazing. In contrast it is the optical-sound image which is really rich when, in the thinness of its description, it gives us no means of making use of the thing from which it derives, instead bringing the thing 'to an essential singularity', one that casts us into a state wherein we no longer know how to react or what to do. It is not just a matter, therefore, Deleuze maintained, of images that are different in kind, but the richness of the optical-sound image in contrast with the poverty of the sensory-motor image.

How can an image which is of no use, which does not engender movement, be called rich? Well, because it is not entirely true that it is of no use. Instead it is a question of type of use. It is true that the optical-sound image does not extend into movement, but that is not to say that it does not go anywhere. Instead it enters into relation with a 'recollection-image'. Indeed, it 'calls up' the recollection-image from the depths of memory. Not only does it have depth, it extends into the depths. It makes itself deep, such that the difference between what is recalled and what is seen and heard in the optical-sound image becomes itself indiscernible (Deleuze, 1989, page 46). Here, in the world of the optical-sound image, 'there are no longer sensory-motor images with their extensions, but much more complex circular links between pure optical and sound images on the one hand, and on the other hand images from time and thought, on planes which all coexist by right' (Deleuze, 1989, page 47). It is in this sense that the optical-sound image is 'useful' for us. It deepens us, taking us down, from the surfaces of movement on which we are otherwise positioned, in our sensory-motored lives. In doing so it makes us a subject in the true sense of the term. For subjectivity itself emerges, not in movement as such, but in the gap between a received and an executed movement. It is not, as Deleuze argues, 'motor or material, but temporal and spiritual'.

Everything which neuroscience assumes about the image, everything which the psychologists of resilience assume about the function of imagination, as well as everything which the social scientists fraught with worry about climate change assume about the image, in enabling the movements through time by which human beings are able to survive the ordeals of the present in order to achieve new and qualitatively more secure conditions of being can be seen to fall within the descriptions of the sensory-motor image. These are images that perform the function and maintain the value of the utility that the bioperfection imperative of our time demands. In contrast, and against this imperative, what our time demands is another quite different image of imagination, one that Deleuze comes close to in his descriptions of the richness and functions of the optical-sound image. For these are images that dislocate us from those spaces wherein the bioperfection imperative may be heard. These are images that may not enable us to survive better but which in their dislocating of us provide us with another image of our self; a self, existing free from the necessity of a life of endless struggle and survival, given over to a temporal and spiritual movement which enriches our existence as free human beings; subjects in the true sense of the term.

Notes

1 Consider also, in the current security climate, here in 2016, the hysteria surrounding the threats posed by images created by Islamic State (ISIS) and disseminated on social media. A wide range of Western security analysts have demanded more attention and greater responses to the functions of these images in the war against terror. See especially Harmanşah (2015, pages 170–7); Giroux (2014).
2 Much of the energy of critical theory in the discipline of International Relations has been devoted to shoring up this central premise of Hobbesian thinking concerning relations between imagination, fear, security and state power. For a now classic account see Campbell (1992).
3 'Resilience thereby comes to be a fundamental mechanism for policing the political imagination, nothing less than the attempted colonisation of the political imagination by the state' (Neocleous, 2013, page 4).

References

9/11 Commission report: the full final report of the National Commission on Terrorist Attacks upon the United States (2004). New York: W. W. Norton and Co.
Butler, J. and Athanasiou, A. (2013). *Dispossession: the performative in the political.* Oxford: Polity.
Campbell, D. (1992). *Writing security.* Minneapolis, MN: University of Minnesota Press.
Chandler, D. (2013). *Resilience: the governance of complexity.* London: Routledge.
Cyrulnik, B. (2009). *Resilience: how your inner strength can set you free from the past.* London: Penguin.
Deleuze, G. (1989). *Cinema 2: the time-image.* London: Athlone Press.
Delsarte, F. (2011). *Delsarte system of expression.* Milton Keynes: Lightning Source.
Evans, B. and Reid, J. (2013). *Resilient life: the art of living dangerously.* Oxford: Polity.
Frith, C. (2007). *Making up the mind: how the brain creates our mental world.* Oxford: Blackwell.

Giroux, H.A. (2014). 'ISIS and the spectacle of terrorism: resisting mainstream workstations of fear'. Available at: https://philosophersforchange.org/2014/10/07/isis-and-the-spectacle-of-terrorism-resisting-mainstream-workstations-of-fear/

Harmanşah, Ö. (2015). 'ISIS, heritage, and the spectacles of destruction in the global media'. *Near Eastern Archaeology*, 78(3): 170–177.

Hobbes, T. (1993). *Leviathan*. London: Penguin.

Hollnagel, E., Woods, D.D. and Leveson, N. (eds) (2006). *Resilience engineering: concepts and precepts*. Aldershot: Ashgate.

Morton, A. (2013). *Emotion and imagination*. Oxford: Polity.

Neocleous, M. (2013). 'Resisting resilience'. *Radical Philosophy*, 178(6): 2–7.

Neocleous, M. (2016). *The universal adversary*. London: Routledge.

Rice, R. (2012). *The Jarmusch way: spirituality and imagination in* Dead Man, Ghost Dog *and* The Limits of Control. Lanham, MD: Scarecrow Press.

Slaby, J. (2015). 'Critical neuroscience meets medical humanities'. *Journal of Medical Humanities*, 41.

Yusoff, K. and Gabrys, J. (2011). 'Climate change and the imagination'. *Wiley Interdisciplinary Reviews*, 2(4): 516–534.

Zuboff, S.Z. (1988). *In the age of the smart machine*. New York: Basic Books.

3

DESIGNING 'SMART' BODIES
Molecular Manipulation as a Resilience-Building Strategy

Rebecca J. Hester

> Biology is the true science of security.
>
> *(Minton, 2017)*

Designing Smart Bodies

Almost overnight, it seems things are becoming smarter: smart cities, smart phones, smart clothes, smart cars and smart home appliances abound. Labelled 'smart' because of their ability to protect themselves, instigate a self-repair or adapt their function to a particular situation or environment (Worden et al., 2003), self-sensing, self-optimising, self-controlling, self-aware and even self-destructing technologies and technological systems hold wide appeal. One of the benefits of smart technologies is their ability to sense changes in their circumstances and to execute measures to enhance their functionality under new circumstances in order to improve performance, efficiency, operating costs and endurance (ibid.). In other words, their promise lies in the fact that their internal capacities are intentionally designed to respond and adapt to external cues in their environment, while at the same time responding to the economic, social and political imperatives programmed into their technology.

Smart technologies are most often conceptualised in terms of hardware and material infrastructures made interoperable through internet communications technology or 'the internet of things'; and made sustainable through engineered adaptability. One of the primary purposes of this interoperability is to make networked systems and environments resilient in the face of vulnerability, danger, catastrophe or disaster.[1] Dexterity, efficiency, plasticity and speed are all characteristics of a resilient system, while the elastic capacity to rapidly adapt and recover characterises the resilient environment (Holling, 1973). One of the key characteristics of smart technologies is their capacity for self-organisation, that is, their inherent ability for iterative adaptation in relation to contextual factors. Whether navigating complexity in the present or returning to a prior form after a disaster, the in-built capacity to weather dynamic circumstances is what makes smart technologies resilient.

Much has been written about the politics of making built infrastructures, like cities, buildings and bridges, smarter and more resilient, while less has been said about the drift in 'smart' thinking toward optimising and enhancing soft infrastructures, like bodies, biology and biological systems. Coining the term *smart body*, this chapter discusses contemporary efforts to design 'smarter' and more resilient living systems through molecular manipulation and modification. Drawing from the literature on resilience as governance, the chapter theorises *smart bodies* and illuminates the politics at work in these nature-culture achievements. Specifically, the chapter is concerned with the ways that biotechnologies are increasingly able to modify our internal environment and, in so doing, to conduct the biological conduct of living matter in order to achieve both scientific and political ends under the guise of resilience, among them increased biological control, improved human health and biotechnological 'fixes' in the face of human-induced biospheric destruction. To the extent that such biotechnological endeavours succeed in designing resilience into the fabric of life itself, the development of *smart bodies* signals molecular-level biopolitical strategies for managing complex life in the present and into the future, while illuminating a molecular dimension of what this volume refers to as 'the resilience machine'.

For the purposes of this discussion, a *smart body* is one whose biology has been modified, either directly or through the creation of environmental incentives, in order to achieve a particular, sometimes heritable, behavioural change at the cellular level with the intent of cultivating an intended relationship between and within biological matter. In other words, a *smart body* is one whose auto-poetic nature has been techno-scientifically sculpted so that it will self-organise in particular ways in relation to both its internal and external environments. Smart bodies thus embody paradoxical logics. On the one hand, their behaviours are contingent, aleatory and self-organising. On the other hand, their self-organisation is guided by the introduction of what Alexander Galloway and Eugene Thacker (2007) refer to as 'protological control'. As detailed later, protological control is a form of embedded governmental control that manages networked behaviour through a diffused rule based on an evolutionary rationale (Galloway and Thacker, 2007, page 61). Through protological control, human imperatives are literally self-propagated in and throughout the tissue of life.

Like other smart technologies, *smart bodies* are modified and managed with an ecological perspective in mind, a perspective that seeks to mimic nature both by taking a holistic and interactive approach to living systems, and by copying the functions and processes of living matter. Biomimicry, or the imitation of nature's patterns and strategies, is thus a key characteristic of *smart bodies*. This approach is not new in biotechnology. Indeed, in his history of biotechnology Bud notes that as far back as the 1960s, genetic scientists prophesised that 'lessons learned from living things can be applied in artificial devices of diverse kinds' (Bud, 1994, page 170). What is novel is that the design of *smart bodies* is also based on technomimicry.[2] That is, the lessons learned from artificial devices and intelligent systems, such as the smart technologies cited previously, are now being applied back into living things. Ideas about the built environment, smart infrastructures and intelligent design are brought to bear on molecular structures through genetic engineering and utilised to modify and manage complex living systems. These nature-culture (Haraway, 2003; Latimer and Miele, 2013) innovations offer both promise and peril.

Informed by current fears of biological catastrophe, smart bodies are designed with the primary objective of protecting human health and security. Indeed, in a context where naturally, intentionally or accidentally occurring biological dangers are held to threaten both critical institutional and economic infrastructures and living systems, there are strong incentives to find ways not just to make life live, but to make it live better and stronger. Pressing global health needs, the devastation wrought by invasive species, as well as climate-induced ecological destruction provide more than sufficient cause to explore the potential of *smart bodies*.

Nevertheless, in using biotechnology to achieve unprecedented levels of biological resilience, there is a danger that such efforts will exceed human control and will literally take on a life of their own. Further, turning to biotechnology rather than to political and social solutions to address existential biothreats serves to solidify the growing perception that life itself is dangerous to living, thus leading to the conclusion that it is biological life itself that must be intervened upon while leaving destructive social, political and economic systems in place. The emergence of *smart bodies* therefore represents efforts to embed resilience within living systems themselves as a response to the perceived threat that life poses to life itself. Because of its focus on health and longevity, biomedical science is at the forefront of these biopolitical efforts. A discussion of *smart bodies* thus illuminates the role of the life sciences and bioengineering in governing life itself through resilience-building strategies, while at the same time adding important dimensions to discussions of contemporary warfare and global health security, the latter of which takes the mitigation of biological danger as its *sine qua non*.

In what follows, and building on the idea that networked circuitry is the defining ontology of the moment, I theorise the *smart body* as a networked life form, the governance of which relies on its capacity for self-organisation and adaptation and on its inherent complexity. Moving beyond much resilience scholarship that foregrounds the neoliberal individual and that focuses on human life, this definition of *smart bodies* operates from the premise that a body is always multiple (Mol, 2002), complex and entangled. Such a conceptualisation incorporates lessons from the 'non-human turn' in the social sciences and feminist scholarship on the body, especially recent work on 'new' or vital materialism, which understands the body as an assemblage that operates ontologically and materially beyond the strictures of any bounded and static individual actor. It also incorporates views from the life sciences, notably the 'molecular vision of life' (Kay, 2000). Within these combined frameworks, the *smart body* is understood as a networked life form that is 'in-formed' by a mix of evolutionary history, human design and contingent and complex relations with other matter. It is this networked and informational understanding of life itself that facilitates both the auto-poetic freedom and the political subjection of *smart bodies*.

I then offer contemporary examples of *smart bodies* and show how protological control facilitates their management. Using evolutionary rationales, scientists seek to harness and modulate the very nature of biological life so that it will evolve and grow in intended ways. Given this, I argue that techno-scientific advancements in biomedical engineering and biotechnology, which seek to make biological matter stronger, healthier and more resilient, enable new, embedded forms of governance within the materiality of life itself. This novel form of governance is meant to facilitate increased biological security through resilience building. Yet, as the final section argues, the development of smart bodies and the politics of ever increasing biological sovereignty also engender their own forms of bio-insecurity (Ahuja, 2016).

Theorising Smart Bodies

Feminist scholars have long debunked the idea that the body is a natural, organic material. Rather, the idea forwarded within much feminist scholarship is that the body is dynamic, uncertain, flexible and volatile (Martin, 1994; Grosz, 1994). Within this perspective, the body is understood to be a complex, emergent and manipulable assemblage of properties and forces that intra-act both ontologically and materially (Barad, 2007). Inspired by such diverse fields as quantum physics, molecular biology, complexity theory and cybernetics, this perspective continues to foreground the matter of bodies – their physical properties – but moves away from reductionistic and mechanistic assumptions of that matter. What we see, instead, is an understanding of biological matter as vibrant and intra-active (Barad, 2007), producing effects and affects within a broader ecology of living things (Bennett, 2010).

For feminist scholars, this process-oriented political ontology offers liberatory possibilities in so far as such theorisations challenge the highly gendered dualistic thinking of Western philosophy. Namely, they trouble the separation between mind and body in which the former stands for masculine reason, sense and transcendence, while the latter stands for feminine passion, unruliness and inert organicity (Grosz, 1994). Theoretically instilling the body with dynamism and complexity while complicating binary distinctions also challenges humanistic notions of liberal individualism that undergird much of Western political thought. If bodies are entangled ontologically, epistemologically and materially, then the idea of individuality (and personal responsibility, individual agency and self-determination) makes little sense.

Such a relational and enactive theorisation challenges long-standing separations in Western philosophy between the human and the non-human replacing them with a world populated not by active subjects and passive objects, but by lively and essentially interactive materials, by entangled agencies – human and non-human (Bennett, 2015, page 224; Barad, 2007; Haraway, 1985). Within these cyborg theories of the body, individuality is replaced by agencies that do not thrive on competition between bounded actors but rather collaboration, connection and kinship within and among entangled bodies. Such theorisations emphasise both the productivity and resilience of matter, highlighting its inventive capacities (Coole and Frost, 2010, pages 7–8) and freedom.

Unlike the finite frames that posit a neoliberal human subject of resilience (Evans and Reid, 2013) or a biological citizen actively optimising his health (Rose, 2007), new materialist and non-human theorisations posit an entire ecology or a complex network of embodied agencies across scales and species that are both productive and resilient. No longer is the political actor a bounded neoliberal human subject, or even a particular species. Rather, the political 'subject' is multiple, emergent, innovative and adaptable matter. This idea of subjectivity maps onto what Felix Guattari refers to as auto-poetic subjectivation, or self-styling, and it accounts both for living organisms and machines (Braidotti, 2013, page 94).

Theories that challenge the coherence of the body (Halberstam and Livingston, 1995) largely draw from what Kroker (2012) calls a code perspective. 'Never fixed and unchanging, code-perspectives are always subject to random fluctuations, always evolving, always intermediated by other objects, by other code-perspectives' (Kroker, 2012, page 2). Inspired by networked thinking and the idea of cellular life as auto-poetic, or self-reproducing, the idea of life as code, as something that could be transcribed, edited and rewritten, has roots in the 1950s and 1960s with the advance of molecular biology. The field of 'new biology', shaped by information theory, cybernetics, systems analyses, electronic computers and simulation technologies, fundamentally altered the representations of animate and inanimate phenomena leading to what Kay calls a 'molecular vision of life' (1993). Within this post-World War II vision, 'genetic information signified an emergent form of biopower: the material control of life would now be supplemented by the promise of controlling its form and logos, its information' (Kay, 2000, page 3). Taken together, the ideas of entangled vitality and life as code have led to an ontological and material understanding of life in terms of networked circuitry. As Galloway and Thacker assert: 'In short, the increasing integration of cybernetics and biology has resulted in an informatic view of life that is also a view of life as a network' (2007, page 51).

Protological Control

The understanding of life as a network engenders particular kinds of techno-political imaginaries. As Kay asserts, within the molecular vision of life 'the genetic code became the site of life's command and control' (Kay, 2000, page 5). Indeed, what was imagined in the 1950s and 1960s as ultimately being commanded and controlled was nothing less than the origins of life, heredity and natural selection. Signalling advances in both biological and computer science and technology, contemporary theorists Alexander Galloway and Eugene Thacker (2007) have challenged the idea that networked life can be commanded and controlled in top-down fashion today, if it ever could. Rather, they suggest that the ability for top-down control evaporates in the emergent form that the network takes. 'In a sense, the power structures have evolved downward, adopting the strategies and structures of the terrorists and the guerrillas' (page 15).

In contrast to a C3 approach (command, control and communications) from above, they describe the ways that networked life can be reorganised through the exercise of protological control from within, a phenomenon they describe as an embedded form of sovereignty. In the politics of networked life, sovereignty is immanent to the network itself as a result of the protocols, or rules, that both organise and facilitate the existence of the network. They explain: 'protocol is two-fold; it is both an apparatus that facilitates networks and a logic that governs how things are done within that apparatus' (Galloway and Thacker, 2007, page 29). In the broadest sense, 'protocol is a technology that regulates flow, directs netspace, codes relationships, and connects life-forms' (page 30).

Arguing against the ideas prevalent in network science that networks are inherently anarchic and anti-authority, and linking biology and cybernetics, they show how protocological control allows networks to emerge and self-organise, albeit in ways guided and directed by protocol, which operates as both form and function. 'Abstracted into a concept, protocol may be defined as a horizontal, distributed control apparatus that guides both the technical and political formation of computer networks, biological systems, and other media' (Galloway and Thacker, 2007, page 29). Using the properties of the network itself, the protocol is able to embed sovereignty within the system and thus ensure the freedom of movement within the network, while also guiding that movement. As they explain, protocological control is thus less about confinement, discipline and normativity than it is about modulation, distribution and flexibility (page 31).

Within their theorisation, freedom is flow. And this flow is directed toward the survival of the network. Recalling that the purpose of the ARPANET (the precursor to the internet) was to develop a network that would be robust enough to withstand the failure of one or more of its nodes, they explain that the objective of protological control is to deal with contingency in the face of loss (Galloway and Thacker, 2007, page 53). Rather than a hard and fast command, protological control is an iterative and adaptive from of control that allows for alternative paths and flexible interactions. It is a way to manage complexity from a distance while also ensuring some level of security to and within the network. In other words, protological control is a way to ensure the resilience of the network, even if some of its nodes cease to function.

Protological control operates at the level of the ecology of the network, securing its continued function not by impeding its interactions, but rather by encouraging freedom of interaction. Indeed, the more interaction that occurs, the more governmental effects can take hold. As Galloway and Thacker explain, the relation between power and protocol is somewhat inverted, 'the greater the distributed nature of the network, the greater the number of inside-out controls that enable the network to function as a network' (2007, pages 54–5). Such 'positive security' was described by Chris Zebrowski in his article 'The nature of resilience':

> Ensuring the subject is capable of co-evolution with their environment cannot be achieved by structuring the mentality of the subject, but was to proceed by acting on the subject's environment, understood as an incentive structure and thus a condition of possibility for emergent norms and behavior ... Security would have to proceed by exposing the subject more fully to the environment so as to optimize its governmental effects in encouraging innovation and, crucially, adaptation.
>
> *(Zebrowski, 2013, page 169)*

Utilising the ideas put forward by Foucault, we can see that freedom and flow within a network are also forms of political subjection. When applied to living systems, protocological control recognises and allows for the governance of contingent, complex and emergent forms of networked biological life by guiding life's autopoetic freedom. Put differently, protocological control modulates the emergence and evolution of encoded life understood as a network without strictly commanding it. Galloway and Thacker explain: 'network control is unbothered by individuated subjects (subjected subjects). In fact, individuated subjects are the very producers and facilitators of networked control. Express yourself! Output some data! It is how distributed control functions best' (2007, page 41).

The idea of networks that self-organise and express themselves based on their protocological structure is helpful for understanding both what smart bodies are – networked information – and the ways that they 'self'-organise, not necessarily in random or boundless fashion nor in ways that respond directly to top-down commands, but through embedded sovereignty. If, as Brad Evans and Julian Reid have argued, 'Building resilient subjects involves the deliberate disabling of the political habits, tendencies, and capacities of peoples and replacing them with adaptive ones' (Evans and Reid, 2013, page 85), then the emergence of *smart bodies* represents a more extreme and immanent technique for achieving such biopolitical ends. At stake are the ways that molecular life, and thus biological life itself, can increasingly be governed at a distance.

This line of argumentation tracks closely the work of David Chandler (2014) on resilience as the new art of governing complexity. Chandler theorises that resilience is an open-ended form of governance that recognises the creative and self-ordering power of complex life. Based on a politics of uncertainty characterised by the unknowability of social phenomena, resilience thinking rearticulates complex life as the positive promise of transformative possibilities. According to Chandler, resilience governs from within the social body; it is embedded *ontologically* in the complexity of life itself, collapsing governance and the object of governance into one. Taking Chandler's idea a step further, I argue that resilience is also progressively embedded *biologically* at the level of life itself collapsing governance and the object of governance – materiality and subjectivity – into one in order to make inter and intra-active life live in controlled, albeit complex and contingent, ways. The effect of this process is the *smart body*. The next section illuminates how this apparatus of biological sovereignty works.

Molecular Manipulation as a Resilience-Building Strategy

> The thing in Nature as high mystery prized,
> This has our science probed beyond a doubt
> What Nature by slow process organized,
> That have we grasped, and crystallized it out.
> *(Goethe, 1948, cited in Myers, 2008, page 7)*

In a 1999 article in *Time* magazine titled, 'How to build a body part' (Fischman, 1999), we learn of the wonders that are occurring in the 'grow your own organs' era. These wonders include: a machinist in Massachusetts who is using his own cells to grow a new thumb; a teenager born without half of his chest wall who is growing a new cage of bone and cartilage within his chest cavity; bladders, grown from bladder cells in a lab, that have been implanted in dogs and are working; and patches of skin, the first 'tissue-engineered' organ to be approved by the US Food and Drug Administration, that are healing sores and skin ulcers on hundreds of patients across the US. To what do we owe these miraculous and heretofore unseen biological developments?

Part of the answer, we are told, is smart engineering. Using materials such as polymers with pores no wider than a toothbrush bristle, researchers have learned to sculpt scaffolds in shapes into which cells can settle. The other part of the answer, however, is 'just plain cell biology'. As Josh Fischman, the article's author, explains:

> Scientists have discovered that they don't have to teach old cells new tricks; given the right framework and the right nutrients, cells will organize themselves into real tissues as the scaffolds dissolve. 'I'm a great believer in the cells. They're not just lying there, looking stupidly at each other,' says Francois Auger, an infectious-disease specialist and builder of artificial blood vessels at Laval University in Quebec City. 'They will do the work for you if you treat them right.'
>
> *(Fischman, 1999)*

Getting cells to behave in intended ways by providing the appropriate guidance is the objective of tissue engineering, a subfield of regenerative medicine. In this burgeoning field, 'researchers aim to create biomimetic scaffolds to mimic the properties of a native stem cell environment, or niche, to dynamically interact with the entrapped stem cells and direct their response' (Zhang, 2012, page 101). Such bioengineering technology works from the idea that cellular relationships will manifest naturally, albeit in ways guided by the scaffold. They will, in other words, manifest the kinds of 'smart' behaviours suggested by their environmental cues.

Such nature-culture achievements are becoming increasingly commonplace in biomedical research. For example, in one experiment an engineered cardiac muscle from neonatal rats was used to measure the strength of the synthetic tissue under stress. The researchers were able to show that by modifying the environment around and between the tissue cells they could control what happens within the cells. That is, they could direct the auto-poetic 'nature' of cellular behaviour. The language used to describe the stakes in the research is telling, 'Their work suggested that engineering the extracellular space is an effective means of enslaving the cardiac myocyte's ability to self-organise its contractile apparatus to maximise the contractile strength of muscle' (Zhang, 2012, page 101). While the idea of an enslaved cell muscle that is at the same time 'self'-organising seems to be somewhat contradictory, it is precisely this kind of biological 'governing at a distance' that researchers intend. Indeed, the fact that engineered tissue can adapt and grow in unknown and changing circumstances, according to researcher intent, is what makes it so scientifically appealing. Further, the idea that the abilities and interactions of biological matter can be subjugated to researcher intent illuminates the political and economic appeal of this technology.

Such cellular governing at a distance is also imagined in a new treatment for type 1 diabetes offered by researchers at the University of California San Francisco. This novel method proposes to transplant live insulin-producing cells enclosed within a flexible membrane by implanting the membrane into diabetic patients who lack those cells. Explaining that 'cells are the ultimate smart machine', the study's architect, Crystal Nytray, argues that living cells work better than mechanical pumps because they can sense which entities to allow through and which to block (Palca, 2017). The idea is that, as an engineered cellular environment, the implanted membrane would keep the cells 'alive and happy' by allowing insulin and blood sugar to pass across the membrane while cells from the recipient's immune system would be kept out, preventing immune rejection. In this complex, multiscalar and smart nature-culture innovation, transplanted cells are made resilient as a result of their built biological environment and, in turn, humans with type 1 diabetes also flourish.

By guiding cellular behaviour, scientists aim to turn living systems into 'factories of the future' designing cells that can not only deliver but also manufacture drugs, food and materials, and that can even act as diagnostic biosensors (Eisenstein, 2016). In one example of this cellular manufacturing, genetically modified microbes are being used to develop smart drugs and living pills (ibid.). In the article 'Smart bug maker Synlogic nabs $5 million from Gates Foundation', we learn that the company is attempting 'to perfect a method of manufacturing microbes that are programmed to sense a specific disease or infection, secrete a drug to treat it, and then self-destruct when they are done' (Fidler, 2014). These smart drugs would work by taking a live bacterium, like a probiotic, and programming in some specific genetic circuitry that no other drug on the market has today. 'With the genetic circuits carefully calibrated, the bacterium can sense the ever-changing conditions it encounters in the body, and adjust the drug dose based on what's needed in real-time' (Timmerman, 2016). Relating its potential to that of an informational technology with in-built intelligence, biotech journalist Ben Fidler illuminates its appeal:

> Think of its potential therapies as 'smart' drugs; tiny bugs engineered logically, like a computer program designed to respond only to specific disease scenarios, and equipped with various fail-safes to make sure they stop when their job is done ... the presumed benefit would be greater control of safety and effectiveness. A custom-designed drug would spare healthy tissue and only attack if the right conditions are present.
>
> *(Fidler, 2015)*

As this example suggests, and operating from the idea of life as code, scientists are creating complex systems by 'wiring up' genetic parts into circuits resulting in 'various living switches and sophisticated sensors' (Eisenstein, 2016) that both detect and react to contextual factors. These in-built molecular modifications are leading to healthier and more resilient species. For example, 'Martin Fussenegger's group at the Swiss Federal Institute of Technology (ETH) in Zurich has built biomedical sensors that can detect disease-relevant metabolites in the blood and trigger the production of therapeutic compounds. In mice, these biosensors successfully staved off gout and obesity, and treated the skin disease psoriasis' (ibid.).

Such smart biological engineering has come a long way in the era of CRISPR-cas9 and gene drive technologies. Rather than altering the environment between cells, CRISPR (short for Clustered Regularly Interspaced Short Palindromic Repeats) can alter the cell itself. Described as a kind of 'genome surgery' (Young Rohjan, 2014) that uses 'molecular scissors' (Regalado, 2016a) to insert, edit or delete DNA in order to achieve a particular biological end, CRISPR-cas9 is driving innovative applications from basic biology to biotechnology and medicine (Hsu et al., 2014). Mimicking a process found in nature, notably a microbial defence system, and reminiscent of the active neoliberal subject, this technique has been called 'active genetics' because 'the mutation actively spreads itself instead of passively being carried into the next generation' (Fikes, 2016). Bypassing the laws of Mendelian inheritance, and much like the tissue cells described above, genes will do the work themselves given the right cues. As a *New York Times* article explains:

Crispr makes it possible to build molecules that can find a particular sequence of DNA inside a cell. The molecules then snip out the sequence, allowing it to be replaced by a different one. The technique might make it possible to introduce not just a gene engineered to reduce fertility in, say, an invasive weasel, but also the genes for the Crispr molecules themselves. Then the weasel would gene-edit itself.

(Zimmer, 2017)

While the mutation is actively doing the spreading and editing, the development and implementation of molecular tools for achieving systems-level change in living systems signals an attempt to 'sculpt' evolution for human purposes. As MIT's Sculpting Evolution website explains: 'By understanding why systems evolve in the ways they do, we are learning to sculpt the evolutionary process and reliably engineer living systems'.[3] Such 'intentional' or 'constructive' biology aims to manage the kinship, heritability and relationality of biological life in order to make bodies and biological systems stronger, more adaptable and resilient in the face of naturally, accidentally or intentionally occurring threats. Put differently, it aims to insert human intentions, values and political rationalities into the evolution of living systems using gene editing as a form of protological control. Compared to such feats as tissue engineering, which entraps cells and works within a scaffold, CRISPR-cas9 gene drive technology is an advanced form of biological control, one that allows for the freedom of biological interaction both within and across species and across generations in a way that mimics nature itself. The MIT Media Lab's website describes the objective: 'By learning to evolve biological systems to acquire and optimize useful characteristics, our creations can approach the remarkable effectiveness of natural living systems.'[4]

This form of protological control was honed in the laboratory of fly biologists at the University of California, San Diego. Scientists Valentino Gantz and Ethan Bier found that they could manipulate the genetic information of a fruit fly, *Drosophilia Melanogaster*, in order to spread a heritable trait, in this case a yellow phenotype, across generations. In computational language that is hard to miss, Gantz described the process of active genetics as a process of modifying a gene so that it 'cuts the other one and copies it over'.[5] While imagined by Gantz and Bier as having important applications for resilience in terms of helping crop strains grow in suboptimal environments, and building better animal models for studying and treating disease, the most significant application of active genetics to date has been in preventing the spread of vector-borne illness. Rather than spraying for mosquitoes and, thus, directly killing them, the introduction of mutagenic chain reactions, or gene drives, into mosquito genes influences the heritability of the mosquito populations. By 'autocatalytically converting heterozygous to homozygous mutation', Gantz and Bier (2015) showed that instead of being passed down in 50 percent of the offspring, as Mendelian genetics would have it, engineered traits can be passed on to up to 99 percent of its progeny (Hesman Saey, 2015).

The success of this approach is based, on the one hand, on hacking the insect's genetic information in such a way that multiple generations of mosquitoes are affected by the molecular modifications introduced into the first generation. This is what Gantz and Bier did with the fruit fly. By changing the molecular structure of the fruit fly, they were able to change the laws of nature as we have come to know them. In other words, they were able to modify genetic rules through protological control. On the other hand, success is based on exploiting the innate capacity of mosquitoes for reproduction and their biological need for species interaction. Because gene drives function only in sexually reproducing species, and because female mosquitoes feed on the blood of other species in order to nurture their own, the relational and contingent aspects of mosquito interactions are core features of the success of this technology. While the technique makes a precise molecular modification in one generation of one species, it gains increased purchase as it is propagated in and through complex and contingent inter and intra-species interactions across time and space. Significantly, and mirroring the cybernetic logic of the internet, in the case of modified insects it is the mutated genetic network that becomes resilient, not necessarily the insect or 'node' itself, as the mutation is passed down to insect offspring. Such controlled complexity is by design.

By genetically modifying mosquitoes, researchers thus intervene at the level of life itself, both in the contemporary time period and into the future, and through this intervention they determine which lives should be made to live (humans) and which should be allowed or made to die (mosquitoes). For example, through genetic engineering, researchers have been able to produce mosquitoes that either fail to reproduce females or that transmit a 'kill' gene that will ultimately wipe out an entire mosquito population. The purpose of these molecular modifications is to prevent malaria, dengue, or the Zika virus in human populations. The concept of 'self-annihilating mosquitoes' works from the idea that mosquitoes interact in complex and contingent ways with each other and with other life forms – those within them (plasmodium) and those without (humans) (Regalado, 2016a). It also, paradoxically, works from the idea that the complexity and contingency of these biological interactions can be managed, though not necessarily commanded, by altering genetic rules. Within this technology, then, species life – human, insect and protozoan – is understood as encoded information that can be rewritten and thus reconfigured together and separately to achieve both political and scientific objectives.

Gene editing is not only being used as a way to prevent vector-borne illness in order to facilitate human flourishing. It is also being utilised to synthetically restore entire ecosystems in order to make them stronger, healthier and more resilient. Researchers working in (human)-assisted evolution (AE), have developed a method for helping coral reefs survive in the face of anthropogenic factors like climate change and destructive fishing. Unlike genetic modification, which alters the phenotype of a biological organism by directly inserting foreign genetic material into it, AE acts to accelerate naturally occurring evolutionary processes through environment modification. The objective of AE is to induce epigenetically controlled stress tolerance that will be passed on to the next generation. This technology has been developed in response to the fact that, for the last few decades, coral reefs all over the world have been bleaching at unprecedented levels. This is significant as corals provide physical and ecological support for a third of all marine life. Known as a 'keystone species', their health is vital for the wellbeing of countless other species, including humans (Riley, 2016).

Although many reefs have died already, researchers believe that they are not completely helpless. Indeed, given time, they cannot only acclimatise to their stressful conditions, but with the assistance of humans in their evolutionary processes they can form a generation of super-corals that will be more robust to the problems of climate change than natural corals (ibid.). In other words, with time and an epigenetic push, coral reefs can become resilient to the dangers that life poses. Ruth Gates, the director of the Hawaii Institute of Marine Biology, explains the stakes, 'It's really about can we raise the resilience of coral?' (ibid.).

Assisted evolution is one way to do that. Similar to tissue engineering and sculpted evolution, AE creates the environmental conditions for reefs to evolve stronger, better and healthier than before by understanding and guiding their auto-poetic nature. 'We're doing what nature does,' says Gates. 'We're just trying to accelerate it, so that corals can keep up. And we'll try to see how far we can push it because, frankly, we've already pushed the planet to accelerate at the rate at which it is warming' (ibid.). In order to move forward, assisted evolutionists are attempting to harness the very nature of life itself in order to make life live. Summarising her approach to this effort, Gates asserts, 'It's all along the lines of what doesn't kill you makes you stronger' (ibid.).

Taken together, the examples provided illuminate both the science and the politics involved in designing *smart bodies*. By altering the rules within which biological life evolves and grows, researchers are able to exercise sovereignty over life itself without directly commanding and controlling the complex and contingent interactions of the living systems they modify. Instead, and as an example of 'protocol as a materialized functioning of distributed control' (Galloway and Thacker, 2007, page 54), researchers create the conditions that will 'in-form' networked interactions. Such cellular governing at a distance operates from the understanding of molecular life as encoded, networked, complex and chaotic. In fact, it is this understanding that facilitates the exploitation of life's auto-poesis for achieving scientific and political objectives. Significantly, these bio-innovations are not about redress in the face of policy failure or biological death. Rather, they aim to structure the course of the future before another biological catastrophe occurs. As Harvard biologist George Church explained to an audience at MIT's Media Lab regarding CRISPR: 'An argument will be made that the ultimate prevention is that the earlier you go, the better the prevention … I do think it's the ultimate preventive, *if* we get to the point where it's very inexpensive, extremely safe, and very predictable' (Regalado, 2015). While not yet at that point, the examples cited earlier demonstrate that multiple efforts are afoot for pre-emptively modifying biological life as a way to prevent undesired biological futures while fostering others.

The Implications of Smart Bodies: Colonising Our Scientific and Political Imaginations

In a time when humans are increasingly recognised as a destructive ecological and geological force, and yet continue to live as if they were not, it should be no surprise that our efforts have turned to biotechnological solutions to make life itself both smarter and more resilient. These imperatives align with a broader restructuring of rationalities and practices comprising liberal governance to manifest what Zebrowski (2013) calls 'the nature of resilience'. Central to this 'nature' is the liberal idea that 'in life itself is to be found the very secret of its security' (Evans and Reid, 2013, page 87). That secret is resilience and in the 'century of biology' (Venter and Cohen, 2004) it is increasingly understood to be found at the molecular level. As sociologist Nikolas Rose (2009) points out, advanced liberal governance has included a whole array of programmes and regulations to equip subjects with, among others, the skills of self-actualisation. He explains how it works: 'As an autonomizing and pluralizing formula of rule, [advanced liberalism] is dependent upon the proliferation of little regulatory instances across a territory and their multiplication, at a "molecular" level, through the interstices of our present experience' (Rose, 2009, page 160).

Through advances in bioengineering these regulatory instances are literally materialised at the molecular level through the cultivation and promotion of the self-actualising or auto-poetic capacities of life itself. The objective of these molecular manipulations is not only to mitigate present or future harm, and thus to make certain lives more resilient, sustainable and governable; it is also to continue our current ways of living without making significant political or economic modifications. As Andrea Crisanti, a genetic engineer working on GMO mosquitoes at Imperial College London, said in response to using gene drive technology in mosquitoes as a way to eradicate malaria: 'Malaria is a problem of poverty, of instability and a lack of a political will … We are asking the drive to do what we can't do politically or economically' (Regalado, 2016b). The lack of political and economic will to change our behaviours, to rethink our land use and to protect the biosphere from anthropogenic insults all but ensures that the only way forward is to rely on biotechnological 'fixes' that alter the biology of things rather than altering human activities, institutional systems and economic programmes.

Rather than evolving our political and economic processes to address the challenges of the Anthropocene, we have, instead, accelerated the evolution of living systems. In fact, as more of our scientific efforts seek to assist evolution, our politics look to increasingly pre-empt it. Brian Massumi has pointed out that pre-emptive thinking is the operative logic of the War on Terror (2015). This logic responds to the uncertainty of the future, to the 'unknown unknowns' of biology as if life itself were the enemy (Ahuja, 2016). It is this uncertainty within living systems that resilience thinking in bioengineering seeks to respond to and, in so doing, pre-emptively transform the molecular future of life itself. The logic inherent to this thinking is that insecurity is the natural order of things and thus individuals and societies must accept catastrophe as the starting point for comporting themselves toward the future (Evans and Reid, 2013).

The idea of sculpting and accelerating evolution while pre-empting it is not a paradox, but rather a manifestation of long-standing efforts to achieve biological control (Rose, 2007). Indeed, the protological management of life itself at the molecular level foretells a new politics of biological control in a longer history of attempts at social control through human engineering (Kay, 1993). This history, and the eugenic values that helped shape it, infused the inception and early practices of molecular biology and the informational gaze that accompanied them (Kay, 1993, 2000). Today, these values have been taken to a new level in the design of smarter, healthier and more resilient bodies – bodies that are central components in what this volume has called 'the resilience machine'. And while Gitte du Plessis (2017) has convincingly argued that there is a futility to continuous efforts to gain environmental and biological control, this has not stopped our political and scientific imaginaries. In fact, in true neoliberal fashion it has only served to spur the political imagination, as Mark Neocleous has argued. He writes that resilience 'is nothing less than the attempted colonization of the political imagination by the state' (2013, page 4). As the examples above demonstrate, resilience thinking is increasingly colonising our scientific imagination as well.

While making life itself more resilient seems, on the face of it, to be an uncontestable good, there are consequences to pre-emptively altering our biological futures. Even as some harmful life forms are prevented from emerging and interacting with others, gene drive technologies can incite living systems to take on a life of their own. Kevin Esvelt explains how this happens with certain forms of genetic engineering: 'The kind of gene drive that is invasive and self-propagating is in many ways the equivalent of an invasive species' (Zimmer, 2017). Mimicking the characteristics of the species it seeks to eradicate, genetic modification can spread to such a degree that it causes damage to the ecologies where it is introduced, thereby creating deadly networks that are themselves resilient. We've no need to think analogously about computer viruses to understand how this works. Anti-microbial resistance is one form such deadly networks take while pandemic viruses are another. Creating new death-inducing networks in the name of making others resilient is only to further exacerbate the challenge of survival, rather than eradicate it. Aubrey Yee illuminates the issue stating: 'In the case of gene drives and engineered mosquitoes, humans have begun generating sovereignty over synthetically produced life whose death is already pre-programmed into its very existence' (Yee, 2017, page 5). As Yee points out, we are pre-emptively killing life in order to make life live.

While for many the death of some species to save others is a worthwhile compromise given the pressing problems of the day, the potential of this technology is suggestive of serious future trade-offs. Kevin Esvelt, the leader in sculpting evolution at MIT, offers some insight into these stating: 'This is why I hate the malaria problem … it makes the technology so tempting to use' (Regalado, 2016b). His comments, alongside those of other scientists intimately engaged with the uses and implications of gene drive technologies, invite us to reflect upon the kinds of inter and intra-species trade-offs we might be willing to make in the future as a survival strategy. Today it is mosquitoes we seek to rewrite and malaria that we seek to edit out, but tomorrow it might be brown eyes or black skin or other human characteristics and capabilities. We may even want to cut or copy certain populations.

Some have compared gene drive technology to 'the next weapon of mass destruction' and have even raised the spectre of insect terrorism, such as mosquitoes that kill people with a toxin (Regalado, 2015). Echoing the concerns of those that are already imagining entomological warfare as a new tool in the military armamentarium, Yee calls attention to the weaponisation of insects and asks,

> how do we ensure that these technologies are not used increasingly against other nonhuman and eventually human "enemies" as we are pushed to increasing velocity through the perpetual state of emergency of pressures like climate change and the mass migrations, food insecurity, water insecurity and war that are becoming characteristic of the 21st-century?
>
> *(Yee, 2017, page 6)*

Genetic engineering scholar at North Carolina State University, Todd Kuiken (2017), takes it a step further asking if the synthetic biology initiatives of the Defense Advanced Research Projects Agency (DARPA) could not only militarise mosquitoes but the entire environment. Kuiken's concerns are not unfounded. As a key member of the Genetic Biocontrol of Invasive Rodents consortium (GBIRd), Kuiken is well poised to understand the political implications of advancing this technology, particularly in light of the fact that DARPA has given $6.4 million to GBIRd and is the world's largest funder of gene drive research.[6]

The very fact that we have these molecular tools at our disposal forces us to pre-emptively decide on our biopolitical futures. The stakes in this decision can be enormously high as Henry Greely, a law professor and bioethics specialist at Stanford, explains: 'environmental uses are more worrisome than a few modified people. The possibility of remaking the biosphere is enormously significant, and a lot closer to realization' (Regalado, 2015). Whether we are more concerned about modified people or modified environments, these concerns signal the profound implications of the colonisation of our political and scientific imaginations for military purposes, as well as the colonisation of entire ecologies and of life itself, under the guise of making living systems healthier, stronger, smarter and more resilient. Indeed, as Kuiken has argued in relation to funding for gene drive research projects: 'researchers who depend on grants for their research may reorient their projects to fit the narrow aims of these military agencies' (Neslen, 2017).

It is not only the state or the US military that can colonise our imaginations and lives, however. Corporations have increasingly privatised the building blocks of life through promises of increased longevity and better health. They have effectively capitalised on the molecular vision of life. As Aubrey Yee has argued, the belief that life can be understood through the basic code of genetic material, DNA, and that reading this like software code allows us to programme and reprogramme the genetic material to do whatever we want it to do is a convenient assumption that leads to the commodification, patenting and corporate control of life forms (Yee, 2017).

One of the dangers here lies in the fact that *smart bodies* are killing to make live while at the same time facilitating increased state and corporate protological control over those lives that are made to live. The other related, and perhaps more insidious danger, of course, is that we all desire lives that are smart, active, free and, above all, resilient. Who, after all, can be against health? Or survival? Or freedom? It is for this reason that we are so tempted by molecular modification as a resilience-building strategy not just to eradicate illness vectors or invasive species but to promote the health and longevity of all valued life forms. Given this temptation, and as engineering and design increasingly influence the field of biology, it is vital therefore that we not only theorise what's at stake in the emergence of *smart bodies*, but also that we continue to grapple with the necro-politics of resilience.

Notes

1 See, for example, the 'Smart and Resilient Cities Editorial Network' Available at: www.smartresilient.com/.
2 My use of technomimicry differs from that of other uses, notably Crabu (2016).
3 www.sculptingevolution.org/research.
4 www.sculptingevolution.org/research/harnessing-evolution.
5 www.youtube.com/watch?v=DhTYRwsCAvQ.
6 http://genedrivefiles.synbiowatch.org/2017/12/01/us-military-gene-drive-development/#21. See also Neslen (2017).

References

Ahuja, N. (2016). *Bioinsecurities: disease interventions, empire, and the government of species.* Durham, NC: Duke University Press.
Barad, K. (2007). *Meeting the universe halfway: quantum physics and the entanglement of matter and meaning.* Durham, NC: Duke University Press.
Bennett, J. (2010). *Vibrant matter: a political ecology of things.* Durham, NC: Duke University Press.
Bennett, J. (2015). 'Systems and things: on vital materialism and object-oriented philosophy', in R. Grusin (ed.), *The nonhuman turn.* Minneapolis, MN: University of Minnesota Press, pp. 223–240.
Braidotti, R. (2013). *The posthuman.* Cambridge: Polity Press.
Bud, R. (1994). *The uses of life: a history of biotechnology.* Cambridge: Cambridge University Press.
Chandler, D. (2014). 'Beyond neoliberalism: resilience, the new art of governing complexity'. *Resilience,* 2(1): 47–63.

Coole, D. and Frost, S. (2010). 'Introducing the new materialisms', in D. Coole and S. Frost (eds), *New materialisms: ontology, agency, and politics*. Durham, NC: Duke University Press, pp. 1–43.

Crabu, S. (2016). 'Translational biomedicine in action: constructing biomarkers across laboratory and benchside'. *Social Theory and Health*, 14(3): 312–331.

Eisenstein, M. (2016). 'Living factories of the future'. *Nature*, 531(7594): 401–403.

Evans, B. and Reid, J. (2013). 'Dangerously exposed: the life and death of the resilient subject'. *Resilience*, 1(2): 83–98.

Fidler, B. (2014). 'Smart-bug maker Synlogic nabs $5 million from Gates Foundation'. Available at: www.xconomy.com/boston/2014/10/07/smart-bug-maker-synlogic-nabs-5m-from-gates-foundation/

Fidler, B. (2015). 'Ex-Pfizer exec Gutierrez-Ramos to lead Synlogic's smart-bug plan'. *Exome*, 21 May. Available at: www.xconomy.com/boston/2015/05/21/ex-pfizer-exec-gutierrez-ramos-to-lead-synlogics-smart-bug-plan/

Fikes, B.J. (2016). 'UCSD gene drive technology offers life transforming power'. *San Diego Union Tribune*, 3 June. Available at: www.sandiegouniontribune.com/business/biotech/sdut-gene-drive-bier-gantz-ucsd-2016jun03-htmlstory.html

Fischman, J. (1999). 'How to build a body part'. *Time*, 1 March. Available at: http://content.time.com/time/magazine/article/0,9171,20592,00.html

Galloway, A.R. and Thacker, E. (2007). *The exploit: a theory of networks* (Vol. 21). Minneapolis, MN: University of Minnesota Press.

Gantz, V.M. and Bier, E. (2015). 'The mutagenic chain reaction: a method for converting heterozygous to homozygous mutations'. *Science*, 348(6233): 442–444.

Grosz, E.A. (1994). *Volatile bodies: toward a corporeal feminism*. Bloomington, IN: Indiana University Press.

Halberstam, J.M. and Livingston, I. (eds) (1995). *Posthuman bodies*. Bloomington, IN: Indiana University Press.

Haraway, D.J. (1985). *A manifesto for cyborgs: science, technology, and socialist feminism in the 1980s*. San Francisco, CA: Center for Social Research and Education.

Haraway, D.J. (2003). *The companion species manifesto: dogs, people, and significant otherness* (Vol. 1). Chicago, IL: Prickly Paradigm Press.

Hesman Saey, T. (2015). 'Gene drives spread their wings: Crispr brings a powerful genetic tool closer to reality. Are we ready?' *ScienceNews*, 188(12): 16.

Holling, C.S. (1973). 'Resilience and stability of ecological systems'. *Annual Review of Ecology and Systematics*, 4(1): 1–23.

Hsu, P.D., Lander, E.S. and Zhang, F. (2014). 'Development and applications of CRISPR-Cas9 for genome engineering'. *Cell*, 157(6): 1262–1278.

Kay, L.E. (1993). *The molecular vision of life: Caltech, the Rockefeller Foundation, and the rise of the new biology*. Oxford: Oxford University Press.

Kay, L.E. (2000). *Who wrote the book of life? A history of the genetic code*. Palo Alto, CA: Stanford University Press.

Kroker, A. (2012). *Body drift: Butler, Hayles, Haraway* (Vol. 22). Minneapolis, MN: University of Minnesota Press.

Kuiken, T. (2017). 'DARPA's synthetic biology initiative could weaponized the environment'. *Slate*, 3 May. Available at: www.slate.com/articles/technology/future_tense/2017/05/what_happens_if_darpa_uses_synthetic_biology_to_manipulate_mother_nature.html

Latimer, J. and Miele, M. (2013). 'Nature cultures? Science, affect and the non-human'. *Theory, Culture and Society*, 30(7–8): 5–31.

Martin, E. (1994). *Flexible bodies: tracking immunity in American culture from the days of polio to the age of AIDS*. Boston, MA: Beacon Press.
Massumi, B. (2015). '*On to power: war, powers, and the state of perception*. Durham, NC: Duke University Press.
Minton, L. (2017). Can biology show us how to stop hackers? Arizona State University, 7 November. Available at: https://asunow.asu.edu/20171107-qa-can-biology-show-us-how-stop-hackers
Mol, A. (2002). *The body multiple: ontology in medical practice*. Durham, NC: Duke University Press.
Myers, N. (2008). 'Molecular embodiments and the body-work of modeling in protein crystallography'. *Social Studies of Science*, 38(2): 163–199.
Neocleous, M. (2013). 'Resisting resilience'. *Radical Philosophy*, 178(6): 2–7.
Neslen, A. (2017). 'US military agency invests $100 million in genetic extinction technologies'. *Guardian*, 4 December. Available at: www.theguardian.com/science/2017/dec/04/us-military-agency-invests-100m-in-genetic-extinction-technologies
Palca, J. (2017). 'A quest: insulin-releasing implant for type-1 diabetes'. National Public Radio, *All Things Considered*, 6 November. Available at: www.npr.org/programs/all-things-considered/2017/11/06/562278314
du Plessis, G. (2017). 'War machines par excellence: the discrepancy between threat and control in the weaponisation of infectious agents'. *Critical Studies on Security*: 1–17.
Regalado, A. (2015). 'Engineering the perfect baby'. *MIT Technology Review*, 5 March.
Regalado, A. (2016a). 'We have the technology to destroy all Zika mosquitoes'. *MIT Technology Review*, 8 February. Available at: www.technologyreview.com/s/600689/we-have-the-technology-to-destroy-all-zika-mosquitoes/
Regalado, A. (2016b). 'The extinction invention'. *MIT Technology Review*, 13 April.
Riley, A. (2016). 'The women with a controversial plan to save corals'. *BBC News*, 22 March. Available at: www.bbc.com/earth/story/20160322-the-women-with-a-controversial-plan-to-save-corals
Rose, N. (2007). 'Molecular biopolitics, somatic ethics and the spirit of biocapital'. *Social Theory and Health*, 5(1): 3–29.
Rose, N. (2009). *The politics of life itself: biomedicine, power, and subjectivity in the twenty-first century*. Princeton, NJ: Princeton University Press.
Timmerman, L. (2016). 'MIT spinout Synlogic grabs $40 million to program microbes into living drugs'. *Forbes*, 17 February. Available at: www.forbes.com/sites/luketimmerman/2016/02/17/mit-spinout-synlogic-grabs-40m-to-program-microbes-into-living-drugs/#21961ee3f75d
Venter, C. and Cohen, D. (2004). 'The century of biology'. *New Perspectives Quarterly*, 21(4): 73–77.
Worden, K., Bullough, W.A. and Haywood, J. (eds) (2003). *Smart technologies*. Singapore: World Scientific.
Yee, A. (2017). *Designing for death: CRISPr and the illusion of control*. Paper presented at the International Studies Association conference, 22 February, Baltimore, MD.
Young Rohjan, S. (2014). 'Rewriting life: genome surgery'. *MIT Tech Review*, 11 February.
Zebrowski, C. (2013). 'The nature of resilience'. *Resilience*, 1(3): 159–173.
Zhang, G. (2012). 'Biomimicry in biomedical research'. *Organogenesis*, 8: 101–102.
Zimmer, C. (2017). '"Gene drives" are too risky for field trials, scientists say'. *New York Times*, 16 November.

4
ORGANISING COMMUNITY RESILIENCE

Chris Zebrowski and Daniel Sage

Introduction

> Have cities missed the point? What if, for all their focus on crime stats and traffic patterns, on zoning laws and building codes, they've missed the chance to thrive—and to help their residents thrive, too? What if there's a better way to govern?
> *(Irving, 2016)*

So begins a blog on the RAND website entitled *A chance to thrive: what we can learn from one city's effort to transform community*. All across America, community resilience programmes are being introduced that promise to enhance happiness, wellness and resilience by restoring the social fabric underpinning American democracy. Originating within literatures on disaster preparedness and recovery (Norris et al., 2008), the idea of community resilience has quickly spread as a solution to a heterogenous array of 'community-based' problems including climate change (Tompkins and Adger, 2004), gang violence (Shirk et al., 2014), violent crime (Ahmed et al., 2004), childhood bullying (Twemlow and Sacco, 2012), poverty (Cutter et al., 2008) and terrorist radicalisation (Weine et al., 2013). By targeting community relations as both the problem and solution to a wide array of highly complex social and political problems, community resilience programmes undoubtedly tap into palpable currents of nostalgia prominent in American public discourse in order to enhance their appeal. And yet, it is also clear, in both their ambitions and the manner in which they reconceive the role of governance that community resilience programmes aim not simply to restore bygone forms of community, but to realise a new idea of what community can, and perhaps should, be.

This chapter critically investigates the forms of community that are promoted and organised through community resilience programmes. In particular, how the social relations comprising the community are understood, problematised and acted upon by the assemblage of governmental practices comprising community resilience initiatives. To assist in these investigations, two bodies of academic literature are drawn on that, to the authors' knowledge, have not been placed in conversation: resilience studies and organisational studies (OS). Critical studies of resilience have demonstrated an historical affinity between resilience and neoliberalism which continues to shape the rationalities and practices governing resilience programmes (Joseph, 2013; Zebrowski, 2013). However, in understanding neoliberalism in terms of 'autonomisation' and focusing critical studies on the 'subject' of resilience (O'Malley, 2010; Reid, 2012), critical studies of resilience have tended to overlook the way in which resilience is being deployed with the aim of restoring and reinvigorating social relations. To help us to analyse these relations we turn to OS. This multidisciplinary field of research is variously concerned with how processes of organising enact distinct social relations, often institutionalised in formal organisations (Parker et al., 2013a; Tsoukas and Knudsen, 2011). Our engagement with OS research enables us to consider how the ways of organising through which community resilience programmes are enacted, whether hierarchical bureaucracy or empowered self-management, shape the socialites they propagate, not least those often considered neoliberal. Our approach echoes Parker et al. (2013b) in viewing organising as 'politics made durable ... a way of working through the complex ways of being human with other humans and hence a responsibility and possibility for us all' (page 39). However, crucially, the durability of politics through organising cannot be assumed in advance but must be worked upon, even by TINA ('There Is No Alternative') neoliberals. By drawing attention to the ways in which community resilience programmes are organised, not simply legitimised through policy discourse, we thus also open a space to question and test the durability of the confluence of neoliberalism and resilience. And we then ask – how might (community) resilience be organised, rendered durable, differently?

We begin this chapter by critically questioning how and why 'communities' became the referent of resilience strategies. This is followed by an examination of the specific social relations promoted and organised through community resilience programmes through a focused study of post-Katrina recovery in New Orleans. We argue throughout that community resilience programmes are driven by a neoliberal idea of a community as a competitive market where human life is to be valued and secured on the basis of its economic productivity (Davies, 2014). Yet despite the predominance of neoliberal governmentalities, in orienting community resilience programmes in New Orleans and elsewhere, the authors are reluctant to reduce resilience to neoliberalism and vice versa. That is why the final section turns to an examination of how some alternative organisations have reworked and supplemented market managerialist organisational practices to enact different notions of community, solidarity and resilience. Significantly, analyses of these alternative work organisations are suggestive of how resilience might be organised differently. In the conclusion to this chapter, we draw out the implications of these alternative ways of organising for rethinking contestation outside of the binary of having to be either for or against resilience.

Social Organisation

Community resilience has been defined as 'the sustained ability of communities to withstand, adapt to, and recover from adversity' (US Department of Health and Human Services, n.d.). Originating in the field of disaster response and recovery, the idea of 'community resilience' signalled a shift from the traditional focus on the individual and household preparedness to the role of social networks in assisting response and recovery efforts. This shift in focus, from the self-sufficiency of the individual household to the social capital underpinning community self-organisation, enabled the migration of discourses of community resilience from the field of emergency response to their wider application as a solution to a variety of 'community-based' problems including health and wellness, public safety, youth development and environmental sustainability. The idea of community resilience draws attention to the psychological, material, physical and socio-cultural resources which allow particular communities to survive, and even thrive, within an environment marked by constant change and uncertainty (Magis, 2010, page 401). Resilience in this respect is understood to be a function of the richness of connections within and across communities. It is both a natural property of communities and a quality which can be improved and extended through good governance.

Within academic research, community resilience operates as a meeting point for two influential, yet distinct, traditions of resilience research: ecological and psychological resilience literatures (see Davoudi et al., this volume). Operating as the point of convergence for these two distinct twin trajectories of resilience research, studies of community resilience promise to provide insight into how communities can actively develop the capacity to adapt to and thrive within environments characterised by perpetual change and turbulence. Here, community resilience is a function of a number of interrelated factors including social capital; the prevalence of social networks; social inclusion; leadership; equality, and an array of psychological factors, including preparedness, ability to cope with change and learning (Buikstra et al., 2010). 'Community' is thus posited as a natural object, endowed with certain inherent capacities of self-organisation, which must be identified, enabled and encouraged through the exercise of good governance. Governing community resilience in this respect is less a top-down process and more one that requires combining, on the one hand, a sense of agency often articulated in terms of individual and/or community 'empowerment', with community action through self-organisation (Magis, 2010).

By appealing to the self-governing capacities of the community, community resilience resonates with certain conservative currents in American political discourse. From the 1990s, a number of influential books have sought to link a raft of contemporary social and political problems to an erosion in the values and forms of organisation which had formerly underpinned American communities (see Etzioni, 1993; Putnam, 2000). Of course, the idea of an authentic community lost in time which requires restoration is an enduring narrative. According to Joseph, the modern discourse of community 'has been dominated by a theme of loss from its inception' (2002, page 6). Indeed, one could argue that the temporal sequencing of community and society is foundational, both to ideas of modernity and the advent of sociology as an academic discipline (Delanty, 2010, pages 6–7). Inherited from medieval Christian millenarian traditions, wherein salvation was understood as the recovery of an original immanence with the Lord in the establishment of his Kingdom on Earth (Turner, 1969, pages 153–4), 'modernity' was conceptualised in terms of a rupture with an authentic past. Whether it be Durkheim's opposition to mechanical and organic solidarity, Weber's narrative of rationalisation or countless others, community is posited as a lost form of sociability, rooted in authenticity, solidarity, trust and shared values, which has been superseded by 'social' modes organised by capitalist values of individualism and economic profit (Joseph, 2002, page 6).

The community/society dichotomy has also been subject to critical scrutiny. Nancy, for one, argues:

> *Society* was not built on the ruins of a *community*. It emerged from the disappearance or the conservation of something – tribes or empires – perhaps just as unrelated to what we call 'community' as to what we call 'society'. So that community, far from being what society has crushed or lost, is *what happens to us* – question, waiting, event, imperative – *in the wake of society*.
>
> (Nancy, 1991, page 11)

For Nancy, the original community never existed. It is mythic thought. Community instead arises as an experience of absence; something which can be desired but never fulfilled. Yet this affect is productive. The experience of loss associated with community, together with the impossibility of its realisation, acts as a constant impetus for the realisation of community as a political project. 'What this community has "lost" – the immanence and the intimacy of a communion – is lost only in the sense that such a "loss" is constitutive of "community" itself' (Nancy, 1991, page 12).

While recognising that the opposition of community against society has been a recurring theme throughout history, Williams (1973) cautions that we must be sensitive to the particular ways in which this opposition is articulated and deployed in specific political and social settings. For Joseph (2002, page 8), what is peculiar about 20th-century invocations of community is the marked absence of anti-capitalist sentiment. Whereas 19th-century appeals to community yearned for a time prior to the commodification of human relations, 20th-century invocations of community seem not just to downplay the influence of capitalism in engendering the forms of sociality they lament, but champion capitalist values as the means by which community can be realised again. This inflection is clearly discernible in Robert Putnam's *Bowling Alone* (1995, 2000). According to Putnam, the steady decline in civic engagement from the 1950s has eroded the social capital of American communities with consequences for the vibrancy of American democracy. In his book, Putnam mobilises a variety of data to empirically prove the decline of American social capital defined as 'social networks and the norms of reciprocity and trustworthiness that arise from them' (Putnam, 2000, page 16). For Putnam, investments in the social capital underpinning communities is not just an important end in itself, but has positive knock-on effects boosting economic and democratic productivity (Joseph, 2002, pages 12–13). The task for government is clear: to identify and invest in the forms of social capital comprising and enriching community.

Nicholas Rose (1996a) has suggested that the re-emergence of discourses of 'community' from the 1980s signal an important shift in the rationalities of governance operating in advanced liberal societies. While acknowledging the historical salience of the community/social divide (Rose, 1996b, page 332), Rose traces the re-emergence of community as an important political term of art to the 1980s, when a number of varied 'community-based' programmes, including community policing, community care, community education and community development, were first introduced. By identifying communities as the principal target of governmental interventions, these programmes signalled a subtle shift away from the social which had acted as the primary referent of social liberal programmes of governance. Like 'communities' in the 1980s, 'the social' emerged as an important governmental term of art in the late 19th century. Assisted by the birth of modern sociology, 'the social', rather than a timeless form or sociality or even a particular mode of organisation, emerged as 'a particular sector in which quite diverse problems and special cases can be grouped together, a sector comprising specific institutions and an entire body of qualified personnel' (Deleuze, 1979, page ix). 'The social' thus acted as a condition of possibility for the development of a suite of 'social' practices, technologies and programmes. Francois Ewald has shown how social insurance emerged as the private technology of insurance was extended to resolve the social problems of the late 19th and early 20th century. By spreading risks across the body of the nation, not only did social insurance operate as a technology of risk management but also acted as a technology of solidarity (Ewald, 1991, pages 209–10). Rose explains:

> It incarnates social solidarity in collectivizing the management of the individual and collective dangers posed by the economic riskiness of a capricious system of wage labour, and the corporeal riskiness of a body subject to sickness and injury, under the stewardship of a 'social' State. And it enjoins solidarity in that the security of the individual across the vicissitudes of a life history is guaranteed by a mechanism that operates on the basis of what individuals and their families are thought to share by virtue of their common sociality. Social insurance thus establishes new connections and association between 'public' norms and procedures and the fate of individuals in their 'private' economic and personal conduct.
>
> (Rose, 1996b, page 48)

Social insurance, in this context, is more than a technique for managing social risks. It is a means of consolidating and reinforcing the social bonds binding the citizens of the nation (Defert, 1991).

Communities represent a novel plane of problematisation and operation for liberal governance. In contrast to the solidarising technologies of social liberalism, Rose argues that programmes of governance aimed at the community effect a 'new spatialization of government' (Rose 1996b, page 327). Whereas social government was oriented towards fostering the relations of obligation between citizen and state towards the realisation of the monolithic nation, neoliberal policies instead target the multiple, overlapping networks of allegiance and responsibility which constitute different 'communities': 'heterogeneous, plural linking individuals, families and others into contesting cultural assemblies of identities and allegiances' (page 327). But community, according to Rose, 'is not simply the territory of government, but a means of governance: its ties, bonds, forces and affiliations are to be celebrated, nurtured, shaped and instrumentalized in the hope of producing consequences that are desirable for all and for each' (page 335). Breaking from the language of 'social risks' which were the target of social liberal policies and technologies of the welfare state, the emergence of 'community-based' policies from the 1980s reflected the idea that governmental practices had to be more specifically tailored to the particular dynamics and risk profiles of individual communities. 'Government through community requires a variety of strategies for fostering and instrumentalizing the multi-layered planes of allegiance any individual may hold to different communities, be they ethnic, religious, sexual, recreational or otherwise' (page 334). Government here is conducted 'through the activation of individual commitments, energies and choices, through personal morality within a community setting' in contrast to what is viewed as 'centralizing, patronizing and disabling social government' (page 335). The use of 'empowerment technologies', which both Rose (1996b) and Donzelot (1991) trace to leftist critiques of the paternalism and overreliance on expert authority in the welfare state, in turn provides a more efficient and economical form of government wherein 'the beings who were to be governed ... were now conceived as individuals who are to be active in their own government' (Rose, 1996b, page 330).

Rather than diminishing power and enhancing freedom, neoliberalism reorients governance towards the production of active citizens, requiring new modes of expertise and governance whist producing new patterns of inclusion and exclusion.

Within resilience research, neoliberalism has become an important paradigm for theorising the 'resilient subject' (O'Malley, 2010; Reid, 2012). Here, the resilient subject is understood as enterprising and entrepreneurial, responsibilised to act on the individual risks they face. Yet, the focus on the individual subject of resilience has deflected attention from the important ways in which programmes of community resilience are increasingly invested in the relations within and between members of distinct communities. This raises important questions concerning the ways in which community, sociality and solidarity are understood and evaluated within academic and practitioner discourses of community resilience.

To address these issues, the next section therefore turns to examine how discourses of community resilience were enacted in the context of post-Katrina recovery efforts in New Orleans. While the analysis starts from a consideration of how community resilience has been framed in policy discourse, it then focuses upon the specific ways of organising through which community resilience has been enacted. In so doing the extent is considered to which specific organisational practices and ideologies, notably management and managerialism (Klikauer, 2013; Parker, 2002), have supported and subverted the policy discourses surrounding community resilience. Considering community resilience as bound up with ways of organising, not simply as a neoliberal policy discourse, thus opens up a space to consider possibilities that it might be organised differently: these possibilities are the subject of the final section.

Organising Community Resilience

At an estimated cost of $62.3 billion, Hurricane Katrina was the costliest 'natural' disaster in American history (Congleton, 2006, page 6). At least 1464 people were confirmed dead and over a million people were displaced from the Gulf Coast region – the largest mass displacement in US history (Grier, 2005). Yet some were able to see a silver lining. David Brooks wrote in the *New York Times* that Hurricane Katrina 'created as close to a blank slate as we get in human affairs, and given us a chance to rebuild a city that wasn't working' (2005). James Glassman, writing in the *Wall Street Journal*, described New Orleans as 'the most exciting urban opportunity since San Francisco in 1906' (2006, page A13). Prior to Katrina, New Orleans had already been singled out as an exemplar of everything wrong with government social welfare policies (Boyer, 2015, page 224). Neoliberals suggested that generous spending on welfare, education and healthcare had produced high levels of dependency amongst the primarily African American population; an argument which resurfaced in the weeks after Katrina to explain why so many African Americans had failed to evacuate the city (Boyer, 2015, page 224). The conditions left by Hurricane Katrina (including the permanent evacuation of 110,000 mostly African American inhabitants) was thus heralded as a historic opportunity to effect a wholesale redesign of the city. New Orleans was quickly designated the principal experimental site for a range of urban redevelopment schemes promising to enhance its economic viability and resilience (Grier, 2005).

Steeped in a discourse of empowerment and self-sufficiency, discourses of resilience provided a much needed boost to morale in the wake of Hurricane Katrina. In contrast to the demeaning media depictions, which focused on destitute individuals amassing at the Superdome or warning of violent criminals looting the streets (Whitehall and Johnson, 2011), resilience provided a counter-narrative around which the citizens of New Orleans could collectively self-identify. The hopeful and redemptive potential of resilience (cf. Konings, 2015), rather than being ideologically imposed, thus found fertile ground within the hopeless affective atmospheres of post-Katrina New Orleans

(Cave, 2005). But resilience also operated as a new paradigm for understanding a diverse array of persistent problems which had been revealed and exacerbated by the floods. *Resilient New Orleans* describes the city as afflicted by a series of chronic 'stresses', including land subsidence and poor economic, educational and health outcomes among vulnerable populations, which act to compound periodic 'shocks' like that wrought by Hurricane Katrina (City of New Orleans, 2015, pages 10–11). Building resilience, it suggests, is a matter of investing in 'equity': a term which deliberately blurs the distinction between pursuing social equality and positioning communities as an investment opportunity (City of New Orleans, 2013, 2015). *Resilience builder: tools for strengthening disaster resilience in your community* provides further insight into how equity and social capital can be enhanced.[1] Freely available on the RAND corporation website, this community resilience toolkit promises to assist communities to 'identify community needs to guide resilience work plans, evaluate progress, and support the development of resilience over the long-term' (LACCDR, n.d., page 3). Injunctions to perform self-assessments, reflect on potential risks and build on existing skill sets and resource bases clearly casts the ideal community member in entrepreneurial terms. However, the solution to the majority of these problems can only be found at the interstice between established communities. Investing in social capital means forging the connections within and between communities which facilitate the circulation of information, skills and resources and enhance the ability of these networks to engage in forms of bottom-up self-organisation in the event of a crisis. On the one hand, complex social divisions with deep historical roots are reduced to technical problems of engagement which can be easily overcome through careful attention to your 'outreach strategy'. ('Many organizational leaders may require in-person meetings, while social media may be a better way to connect with youth groups' (LACCDR, n.d., page 45)). On the other hand, the emphasis on building equity and investing in social capital reframes the community as an investment opportunity wherein market relations are naturalised as the authentic social bond organising communities, thus opening it to a particular rationality of economic governance.

This economic rationality of governance is also discernible in the recovery strategy outlined in *The Unified New Orleans Plan* (UNOP) (City of New Orleans, 2007). UNOP was presented in the wake of the failure of *The Action Plan to Rebuild New Orleans* (2006) which drew intense criticism for suggesting that large sections of New Orleans – identified by large green dots on the accompanying map – were unsuitable for redevelopment. Campanella (2015, page 104) describes how resistance to the 'green dot plan' led to massive community organisation on the part of citizens of New Orleans who self-organised to resist the demolition of their communities. The failure of the green dot plan in the face of public protest put considerable pressure on the mayor's office who had been instructed by the Federal Emergency Management Agency that a recovery plan needed to be in place before federal reconstruction aid would be made available. UNOP was completed in June 2007. In contrast to previous plans, UNOP was rooted in a discourse of participatory planning understood as 'a mechanism of shared governance, where all residents can participate as collective authors of the city's reconstruction directive' (Barrios, 2011, page 118). Yet contrary to the official representations of local government as a process of 'shared governance', Barrios found that recovery planning was often characterised by 'significant tensions' between expert planners and community members over the trajectory of the reconstruction efforts. Professional planners and architects viewed communities as sites of capital investment where public money was to be directed to projects that promised to attract follow-on private investment. Plans re-envisioned the city as a function of circulations of capital and people with commercial corridors connecting important landmarks. These landmarks included funding for sites, such as the Katrina Memorial and new green spaces, formally designed to target and channel diffusive affects such as community 'hope' and 'strength' around this specific vision for urban renewal (City of New Orleans, 2007, pages 279, 305). However, this vision of urban development clashed with neighbourhood and community members' preference for the restoration of a familiar urban design, and the need for social and affordable housing in order to facilitate the return of residents. Barrios (2011) discusses how tensions were compounded by a distinct lack of meaningful public participation in the planning process behind UNOP: significant planning decisions (e.g. to build green spaces to replace affordable housing) were taken with minimal public involvement, highly developed and aesthetically polished plans were then presented at public forums and the ensuing concerns of public actors were, when they conflicted with the views of planners, often disregarded in finally approved plans including UNOP.

Barrios (2011) frames the ethos of UNOP around the 'neoliberal injunction to subject all aspects of social life to capitalist logics of investment and cost-benefit' (page 124). However, while a neoliberal refrain of recovering urban competitiveness (Davies, 2014) undoubtedly infuses UNOP, the expert-led, top-down organising process by which that refrain was naturalised appears distinctly managerial (Klikauer, 2013). UNOP is articulated through a catalogue of management control: 'leadership', key performance indicators, implementation projects, visions, strategic frameworks and boards of directors. UNOP's foreclosing of 'community' and 'resilience' around economic productivity, entrepreneurship and competition (Barrios, 2011) is as much managerially enabled as it is driven by a neoliberal injunction. As Land (2010) stresses: 'as soon as they are subject to management and functional control ... communities are no longer primarily concerned with the maintenance of community itself or with the "common" interests of the community members' (page 128). That is, it is not neoliberalism per se that instrumentalises notions of 'community resilience' to serve and protect external economic logics, but rather neoliberalism works through already present mechanisms for externally defining, managing and controlling 'community'. In the case of New Orleans, urban managerialism was long established: in the 1960s new highways were built that divided families, just as in the 1970s urban development displaced families (Barrios, 2011). What is distinct about the confluence of neoliberalism and managerialism after Katrina is that it is the capitalist corporation (Bakan, 2005), not a modernist urban utopia, that serves as the external yardstick to value human communities (Davies, 2014). Thus, the loss of a section of the population from New Orleans' Lower Ninth Ward after Katrina (Barrios, 2011) is only regarded as problematic if its loss negatively impacts on measures of urban competitiveness or productivity, not because of any shared notion of community collapse or fragmentation. Communities are to be managed from elsewhere into competitive products to sell to investors in global marketplaces for urban living or else treated, like the Lower Ninth Ward, as places to 'externalise' costs such as urban crime, environmental degradation and social inequality (Bakan, 2005) and be (dis)organised into terminal decline (Barrios, 2011).

Despite the emphasis on inclusion within discourses of community, the managerialist practices through which community resilience was fostered appeared to engender and inflame the very divisions between communities that discourses of community resilience appear so ignorant of in the first place. Explicit forms of exclusion, such as those evident within the 'green dot plan', demonstrate a prioritisation on attracting the entrepreneurial classes by transforming New Orleans into a 'New American City' (BNOB11, 2006; see also Davoudi et al., this volume). Yet, despite the prevalence of neoliberal enframings of community resilience it is important to recognise that these interpretations are not hegemonic. The work of the Institute of Women and Ethnic Studies (IWES) in seeking to 'reframe resiliency'[2] is instructive in this regard. This campaign aims to shed a light on the inequitable and unjust recovery support provided to communities battling racism and poverty for decades prior to Hurricane Katrina. Unlike the RAND toolkits, this work recognises the deep historical roots of political divisions between the communities of New Orleans, including histories of slavery, segregation and racism. Moreover, instead of blaming oppressed groups for their lack of 'social capital', they draw upon the cultural practices which permitted these groups to both survive and resist through long periods of abject oppression. In this way, the work of IWES is interesting in so far as it refuses to 'resist resilience' (Neocleous, 2013; Reid, 2012) and instead insists upon recognising and redeploying the cultural practices and modes of organisation that already underpin the resilience of these communities *as a form of resistance*. The work of IWES in attempting to 'reframe resilience' raises questions about the place of resilience discourses within progressive political programmes. This question is now turned to in the final section.

Alternative Organisations of Resilience

While acknowledging the influence of neoliberal governmentalities in shaping the genealogy of resilience, critical resilience scholars have begun to question the subordination of resilience to neoliberal ideology (Anderson, 2015; Sage et al., 2015; Zebrowski, 2016). In this final section, there is an exploration of some of the ways in which notions of 'community' and 'resilience' are being organised in ways that challenge or explicitly oppose neoliberal ideologies. In order to pursue this line of questioning, the chapter now turns towards work within organisation studies that has examined alternative organisation. This small but growing body of work has developed over the last decade as a response to the recognition that critiques of the market managerialist version of organisation, concomitant with neoliberalism, are shorn of their radical potential if they do not move beyond handwringing criticism (Spicer et al., 2009; Voronov, 2008). For the purposes of this chapter, the research opens up possibilities that market managerialism, the professed organisational counterpart to neoliberal ideology, is not the only response to how to (re)organise communities after events such as Hurricane Katrina.

A recent special issue of *Organization* on worker cooperatives as an organisational alternative was framed by the editors not only as a response to the financial crisis but as a way of organising with, not against, community: 'The narrowing of the vision of strategic planning in many industries, along with increasing pressures within international financial markets, has further distanced the most common forms of capitalism from the concerns of community, including attention to employee welfare, attachment to place, and overall social and environmental progress' (Cheney et al., 2014, page 592).

Here, long-standing nostrums of 'community', including solidarity, shared values, trust and a stronger attachment to place, are framed as the effect of worker cooperatives. Thus, while a distinction is made between communities and worker cooperatives, this distinction is traversed by a shared form of social solidarity that binds together community sites of reproduction, and other non-work activities (e.g. recreation, religion), and those of economic production. The diffusion of social solidarity is claimed to provide resilience, or durability, across worker cooperatives and geographical communities (Cheney et al., 2014). These ideas contrast to how market managerialism has co-opted a discourse of corporations as communities to bind individuals to economic strategies (Land, 2010). Significantly, these corporate treatments of community are premised upon precarity, not resilience or sustainability. By contrast, in worker cooperatives, during times of shock, organisational practices such as hierarchical pay structures may themselves be abandoned in the pursuit of community resilience. Cheney et al. (2014) cite a large number of studies, and cases, linking worker cooperatives to increases in employee motivation and wellbeing and social and environmental responsibility. While these celebrations are not explicitly framed in terms of a discourse of 'community resilience', an explicit, if undertheorised, parlance of 'resilience' and 'community' is used to describe how worker cooperatives can foster solidarity as a protective response to a range of natural and human shocks (page 595).

And yet, despite the evidence deployed by Cheney et al. that worker cooperatives can aid communities to withstand certain shocks, such alternative organisations are continually challenged with the degeneration of their principles of solidarity. For example, Cheney et al. provide the example of how the Mondragon Cooperative Corporation (MCC) when faced with declining sales, and a lack of profitability, in an electronics subsidiary in Spain opted to close this division rather than continue to cross-subsidise it. Nevertheless, two thirds of workers employed at that subsidiary were provided with equivalent employment elsewhere within the group (Cheney et al., 2014, page 594). Moreover, the same cooperative has also on occasion introduced, and then removed, stealth management techniques such as Just-in-time and Total Quality Management associated with work intensification and employee reduction.

MCC originated in 1956 in the Basque Country in Spain, and remains committed to developing the interests of that geographical community and others through its principles of social transformation especially in areas such as education (Heras-Saizarbitoria, 2014; see also Cheney, 1999). However, since the early 1990s MCC has expanded internationally to have a presence in 41 countries, exporting to over 150 and now employing 10,000 people outside Spain in areas as diverse as industry, retail, distribution and finance; thus, it serves as much as a model for corporate social responsibility and ethical management as radical alternative organising (Cheney et al., 2014; Flecha and Ngai, 2014). MCC's managers framed this strategy as 'social expansion', wherein the cooperative values of solidarity and economic democracy were to be exported to their global subsidiaries (Flecha and Ngai, 2014). MCC sought to promote a culture of solidarity by developing practices of employee participation and self-management across many of its global subsidiaries. However, MCC's global expansion has largely occurred through the acquisition, or part-acquisition, of capitalist enterprises. This strategy equips MCC employee-investors from Spain with greater freedom to adapt to particular economic shocks causing losses in a host country (e.g. reducing wages, selling assets, selling the company). In other words, despite some attempts to disseminate collectivisation, here a model of solidarity as resilience in the Basque Country was rendered durable by promoting a model of economic self-responsibility as resilience in another part of the world. Furthermore, both Storey et al. (2014) and Heras-Saizarbitoria (2014) conclude that in recent years the culture of MCC in Spain itself has shifted from communitarian values of solidarity and sacrifice to an individualist culture bounded together by the organisational offer of secure employment in the wake of global financial crises. The extent to which this valorisation of employment security is read as either an individualistic debasement of traditional Basque community values (as with Heras-Saizarbitoria, 2014) or a necessary reworking of ideals of community resilience and solidarity in response to volatile global economy and work precarity is debateable.

These complex, and nuanced, dynamics between neoliberalism, alternative organising and community are examined more explicitly in Meira's (2014) empirical study of worker-led factory takeovers in Brazil. Acknowledging the degeneration of alternative organisations into market managerial enterprises, Meira suggests in detail how alternative organising can promote a strong vision of community under neoliberalism. Central to Meira's thesis is the notion of 'communitas' as liminality developed by the anthropologist Victor Turner. Meira elaborates how factory workers, turned owner-managers, are haunted by their precarious encounter with market managerialism, and neoliberalism during the proposed factory closure. Their past status, and the ensuing collapse of social structures (e.g. division of labour, pay differences, occupations) required to bring the factory back from the brink prevents them from fully subscribing to the structure of market management even as they collectively adopt many practices associated with it (e.g. decision hierarchies, limited company status, outsourcing, consultant hiring, Total Quality Management). For example, the organisation remains committed to collective decision making amongst the 27 factory owners (though not the wider employee group) and so frequently struggles to pursue profitability over solidarity. Liminality is, for Meira, a useful basis for thinking how community, for solidarity and sacrifice, can be organised in a neoliberal age. Community solidarity and sacrifice are not engendered here through alternative organising that works with a revolutionary, utopian zeal to produce a better social structure (2014, page 717). Rather, alternative organising fosters solidarity through organisational liminality, an anti-structure (communitas) that arises in the gap between different social structures (in this case a liminal region between a shareholder-owned factory and a full worker cooperative). Viewed as such, alternative organisation incubates community within the market managerialist structures of neoliberalism: 'Communitas is in charge of the company' (page 725).

In this section, it has been considered how a trajectory of work on alternative organisation might help wrestle treatments of community resilience away from their neoliberal designation. The decision to focus here upon alternative work organisations, specifically worker cooperatives, including those closely resembling capitalist enterprises such as MCC, draws attention to how such sites and practices of alternative organising offer an important means to understand, and render durable, different articulations of solidarity, sacrifice, economic democracy and above all perhaps community resilience. These socialities contrast with how community resilience has, as in the case of Hurricane Katrina, often been prefigured in terms of individual empowerment, entrepreneurialism and competition. However, significantly, these socialities are shown to be organised within, not outside or even in direct opposition to, sites often celebrated as neoliberal exemplars, such as retail enterprises. But unlike with corporations, 'community' is not purely figured as a place to externalise costs (Bakan, 2005, pages 60–84) or speciously instrumentalised as a synonym for corporate culture (Land, 2010, pages 114–29) but rather 'community' is defined, enacted and rendered resilient, through processes of organising (Parker et al., 2013a).

Conclusion

In the conclusion of *The Death of the Social*, Rose (1996b, page 328) warns progressives 'against holding too sentimental an attachment to 'the social'. 'The social,' he reminds us, 'is invented by history and cathected by political passions: we should be wary of embracing it as an inevitable horizon for our thought or standard for our evaluations' (page 329). Whilst recognising the investment of socialism into the numerous social instruments comprising the welfare state – social insurance, social protection, social services, etc. – Rose nevertheless questions the political expediency of reacting to displacement of the social by the community by holding strictly to a discourse of the social, that was itself not without its problems. 'We need not simply to condemn the injustices and disadvantages entailed in the de-socialization of government, but also to engage inventively with the possibilities opened up by the imperatives of activity and the images of plural affinities' (page 358).

It is clear that neoliberal governmentalities predominantly enframe the ways in which community resilience is enacted through programmes being introduced across the United States. Circumscribed within a neoliberal rationality of governance, community resilience programmes aim to bolster the adaptive, self-organisational capacities of communities by enriching social capital and promoting entrepreneurial forms of behaviour. The understanding of community is thus reinterpreted through market managerial and economic discourses. Community relations are, in turn, understood as a form of capital requiring sustained investment while market relations are naturalised as the authentic basis of human sociality. Yet, we have been reluctant to conflate resilience with neoliberalism. By turning to the experience of alternative work organisations, we have sought to open a line of questioning on how community, solidarity and resilience might be enacted in a manner resistant to the logics and practices of contemporary neoliberalism. What is clear from our investigation is the difficulty of establishing modes of organisation entirely separate from neoliberal governmental logics they seek to oppose. It is for this reason that Meira describes the positionality of workers in terms of their liminality: seeking out opportunities for resistance by resorting to 'a sort of creativity by necessity that shadows the system more than confronts it' (Meira, 2014, page 714). Interestingly, Meira frames resistance in terms of ethical uncertainty. In seeking to contest the demands of transnational capital, one must always be wary of, on the one hand, reproducing and/or reaffirming the logics one is seeking to contest, and on the other, of slipping back into a form of isolationism. However, rather than weakening their resolve, Meira sees this ongoing liminal organising as constitutive of the community itself.

In line with Rose's plea for an engaged form of critique, that moves beyond dismissal in order to explore the creative opportunities presented by the rise of communities as a post-social object of governance we feel compelled, as critical scholars, to explore the progressive opportunities afforded by resilience. Rather than simply 'resisting resilience' we are curious as to how critical resilience studies might attend to the ways in which resilience ideas are being creatively reworked in ways which contest market managerialism and wider programmes of neoliberal governance. This may be less a matter of defining what resilience 'is' than exploring the possibilities harboured within resilience discourses to foster experimentation on what it could signify and how it could be made durable in different ways.

Clearly this is both difficult and dangerous: running the risk of reaffirming practices and logics ultimately sought to be problematised. However, the authors feel that critical resilience studies should not shy away from asking difficult questions concerning the potential of resilience discourses to reinvigorate a progressive alternative to neoliberal governmentalities. How can resilience be enacted in ways resistant to the rationalities and practices of neoliberalism? How might resilience be reinterpreted and differentially to organise and render durable novel forms of community? What would a resilient community look like? What dangers are there in romanticising community resilience? Enjoining such a line of questioning to alternative organisational practices may provide important insights on how to move resilience beyond its restrictive neoliberal enframing.

Notes

1 www.rand.org/multi/resilience-in-action/community-resilience-toolkits.html.
2 See http://www.iwes-resilience.org/

References

Ahmed, R., Seedat, M., Van Niekerk, A. and Bulbulia, S. (2004). 'Discerning community resilience in disadvantaged communities in the context of violence and injury prevention'. *South African Journal of Psychology*, 34(3): 386–408.
Anderson, B. (2015). 'What kind of thing is resilience?'. *Politics*, 35(1): 60–66.
Bakan, J. (2005). *The corporation: the pathological pursuit of profit and power*. London: Constable and Robinson.
Barrios, R.E. (2011). '"If you did not grow up here, you cannot appreciate living here": neoliberalism, space-time, and affect in post-Katrina recovery planning'. *Human Organization*, 70(2): 118–127.
BNOB11 (2006). *Action plan for New Orleans: the new American city*. New Orleans: Bring New Orleans Back Commission, 11 January.
Boyer, M.C. (2015). 'Katrina effect: the ruination of New Orleans and the planners of injustice', in W.M. Taylor, M.P. Levine, O. Rooksby and J.-K. Sobott (eds), *The "Katrina effect": on the nature of catastrophe*. London: Bloomsbury, pp. 223–254.
Brooks, D. (2005). 'Katrina's silver lining'. *New York Times*. Available at: www.nytimes.com/2005/09/08/opinion/katrinas-silver-lining.html?_r=0
Buikstra, E., Ross, H., King, C.A., Baker, P.G., Hegney, D., McLachlan, K. and Rogers-Clark, C. (2010). 'The components of resilience: perceptions of an Australian rural community'. *Journal of Community Psychology*, 38(8): 975–991.
Campanella, R. (2015). 'A Katrina lexicon', in W.M. Taylor, M.P. Levine, O. Rooksby and J.-K. Sobott (eds), *The "Katrina effect": on the nature of catastrophe*. London: Bloomsbury, pp. 91–114.
Cave, M. (2005). 'Through hell and high water: New Orleans, August 29–September 15, 2005'. *Oral History Review*, 35(1): 1–10.
Cheney, G. (1999). *Values at work: employee participation meets market pressure at Mondragon*. Ithaca, NY: Cornell University Press.

Cheney, G.Cruz, I., Peredo, A. and Nazareno, E. (2014). 'Worker cooperatives as an organizational alternative: challenges, achievements and promise in business governance and ownership'. *Organization*, 21(5): 591–603.
City of New Orleans (2007). *The unified New Orleans plan: citywide strategic recovery and rebuilding plan (UNOP)*. New Orleans.
City of New Orleans (2013). *Prosperity NOLA: a plan to drive economic growth for 2018*. New Orleans.
City of New Orleans (2015). *Resilient New Orleans: strategic actions to shape our future city*. New Orleans.
Congleton, R.D. (2006). 'The story of Katrina: New Orleans and the political economy of catastrophe'. *Public Choice*, 127(1–2): 5–30.
Cutter, S.L., Barnes, L., Berry, M., Burton, C., Evans, E., Tate, E. and Webb, J. (2008). 'A place-based model for understanding community resilience to natural disasters'. *Global Environmental Change*, 18(4): 598–606.
Davies, W. (2014). *The limits of neoliberalism: authority, sovereignty and the logic of competition*. London: SAGE.
Defert, D. (1991). '"Popular life" and insurance technology', in G. Burchell, C. Gordon and P. Miller (eds), *The Foucault effect: studies in governmentality*. Hemel Hempstead: Harvester Wheatsheaf, pp. 211–234.
Delanty, G. (2010). *Community* (2nd edn). London: Routledge.
Deleuze, G. (1979). 'Forward: the rise of the social', in J. Donzelot (ed.), *The policing of families*. New York: Pantheon Books, pp. ix–xix.
Donzelot, J. (1991). 'The mobilization of society', in G. Burchell, C. Gordon, and P. Miller (eds), *The Foucault effect: studies in governmentality*. Hemel Hempstead: Harvester Wheatsheaf, pp. 169–180.
Etzioni, A. (1993). *The spirit of community: rights, responsibilities, and the communitarian*. New York: Crown Publishers.
Ewald, F. (1991). 'Insurance and risk', in G. Burchell, C. Gordon and P. Miller (eds), *The Foucault effect: studies in governmentality*. London: Harvester Wheatsheaf, pp. 197–210.
Flecha, R. and Ngai, R. (2014). 'The challenge for Mondragon: searching for the cooperative values in times of internationalization'. *Organization*, 21(5): 666–682.
Glassman, J.K. (2006). 'Cross country: back to the future'. *Wall Street Journal*, A13.
Grier, P. (2005) 'The Great Katrina Migration'. *Christian Science Monitor*, 12 September. Available at: www.csmonitor.com/2005/0912/p01s01-ussc.html
Heras-Saizarbitoria, H. (2014). 'The ties that bind? Exploring the basic principles of worker-owned organizations in practice'. *Organization*, 21(5): 645–665.
Irving, D. (2016). 'A chance to thrive: what we can learn from one city's effort to transform community'. RAND Blog. Available at www.rand.org/blog/rand-review/2016/03/a-chance-to-thrive.html
Joseph, M. (2002). *Against the romance of community*. Minneapolis, MN: University of Minnesota Press.
Joseph, J. (2013). 'Resilience as embedded neoliberalism: a governmentality approach'. *Resilience: International Policies, Practices and Discourses*, 1(1): 37–41.
Klikauer, T. (2013). *Managerialism: a critique of ideology*. Basingstoke: Palgrave.
Konings, M. (2015). *The emotional logic of Capitalism: what progressives have missed*. Stanford, CA: Stanford University Press.
LACCDR (Los Angeles County Community Disaster Resilience Project) (n.d.). *Resilience builder: Tools for strengthening disaster resilience in your community*. Los Angeles, CA.

Land, C. (2010). 'Community', in P. Hancock and A. Spicer (eds), *Understanding corporate life*. London: SAGE, pp. 114–129.

Magis, K. (2010). 'Community resilience: an indicator of social sustainability'. *Society and Natural Resources: An International Journal*, 23(5): 401–416. Available at: http://doi.org/10.1080/08941920903305674

Meira, F. (2014). 'Liminal organization: organizational emergence within solidary economy in Brazil'. *Organization*, 21(5): 713–729.

Nancy, J.-L. (1991). *The inoperative community*. Minneapolis, MN: University of Minnesota Press.

Neocleous, M. (2013). 'Resisting resilience'. *Radical Philosophy*, 178(6): 2–7.

O'Malley, P. (2010). 'Resilient subjects: uncertainty, warfare and liberalism'. *Economy and Society*, 39(4): 488–509.

Parker, M. (2002). *Against management*. Cambridge: Polity Press.

Parker, M., Cheney, G., Fournier, V. and Land, C. (eds) (2013a). *The Routledge companion to alternative organization*. London: Routledge.

Parker, M., Cheney, G., Fournier, V., Land, C. and Lightfoot, G. (2013b). 'Imagining alternatives', in M. Parker, G. Cheney, V. Fournier and C. Land (eds), *The Routledge companion to alternative organization*. London: Routledge, pp. 31–41.

Putnam, R.D. (2000). *Bowling alone: the collapse and revival of the American community*. New York: Simon & Schuster.

Reid, J. (2012). 'The disastrous and politically debased subject of resilience'. *Development Dialogue*, 58: 67–80.

Rose, N. (1996a). 'Governing "advanced" liberal democracies', in A. Barry, T. Osborne and N. Rose (eds), *Foucault and political reason: liberalism, neo-liberalism, and rationalities of government*. Chicago, IL: University of Chicago Press, pp. 37–64.

Rose, N. (1996b). 'The death of the social? Re-figuring the territory of government'. *Economy and Society*, 25(3): 327–356.

Sage, D., Fussey, P. and Dainty, A. (2015). 'Securing and scaling resilient futures: neoliberalization, infrastructure, and topologies of power'. *Environment and Planning D: Society and Space*, 33(3): 494–511.

Shirk, D.A., Wood, D. and Olson, E.L. (eds) (2014). *Building resilient communities in Mexico: civic responses to crime and violence*. Washington, DC: Wilson Centre Mexico Institute.

Spicer, A., Alvesson, M. and Kärreman, D. (2009). 'Critical performativity: the unfinished business of critical management studies'. *Human Relations*, 62(4): 537–560.

Storey, J., Basterretxea, I. and Salaman, G. (2014). 'Managing and resisting "degeneration" in employee owned businesses: a comparative study of two large retailers in Spain and the United Kingdom'. *Organization*, 21(5): 626–644.

Tompkins, E.L. and Adger, W.N. (2004). 'Does adaptive management of natural resources enhance resilience to climate change?'. *Ecology and Society*, 9(2).

Tsoukas, H. and Knudsen, C. (eds) (2011). *The Oxford handbook of organization theory*. Oxford: Oxford University Press.

Turner, V. (1969). *The ritual process: structure and anti-structure*. Ithaca, NY: Cornell University Press.

Twemlow, S.W. and Sacco, F.C. (2012). *Preventing bullying and school violence*. Washington, DC: American Psychiatric Publishing.

US Department of Health and Human Services (n.d.). 'Community resilience'. Public Health Emergency website. Available at: www.phe.gov/Preparedness/planning/abc/Pages/community-resilience.aspx

Voronov, M. (2008). 'Toward engaged critical management studies'. *Organization*, 15(6): 939–945.

Weine, S., Henderson, S., Shanfield, S., Legha, R. and Post, J. (2013). 'Building community resilience to counter violent extremism'. *Democracy and Security*, 9(4): 327–333.

Whitehall, G. and Johnson, C. (2011). 'Katrina refugees and neoliberal self-governance', in C. Johnson (ed.), *The neoliberal deluge: Hurricane Katrina, late capitalist culture and the remaking of New Orleans*. Minneapolis, MN: University of Minnesota Press, pp. 60–86.

Williams, R. (1973). *The country and the city*. New York: Oxford University Press.

Zebrowski, C. (2013). 'The nature of resilience'. *Resilience: International Policies, Practices and Discourses*, 1(3): 159–173.

Zebrowski, C. (2016). *The value of resilience: securing life in the 21st century*. London: Routledge.

5

REJECTING AND RECREATING RESILIENCE AFTER DISASTER

Raven Cretney

Introduction

Resilience has risen to prominence within the context of increasingly diverse and contextualised neoliberal forms of governance and governmentality (Cote and Nightingale, 2011; Neocleous, 2013; Walker and Cooper, 2011). It is integral, however, to expand and explore the ways in which resilience is also enacted and envisioned as a radical and transformative alternative to that which has emerged in this normative context. The origins of resilience thought, whether in human society or ecosystems, speak to a desire to shift, adapt and transform the organisation of space and function in response to disruptions to a perceived status quo (Adger, 2000; Gunderson, 2010). While these notions have been rightfully critiqued, they also hold the potential for exploring the many different and varied ways communities are engaging with the idea.

As has been demonstrated in the previous chapters of this book, the machine-like enactment of resilience by a range of actors, across an array of institutions and aspects of social and political life, can serve to reinforce and extend ideologies of neoliberal capitalism. This pervasive and extensive approach to colonising the political, social and economic sphere has manifested in diverse and contextualised forms, particularly at different scales. The everyday scale is a site in which these discourses of self and society can be internalised and normalised through the rhetoric of resilience. However, as will be argued in this chapter, the machinic assemblage of resilience produces not only the power to reinforce but also the power to contest and resist.

Drawing on a machinic framing of resilience, it is possible to conceptualise how the resilience machine can both reinforce the status quo of neoliberal capitalism while also holding space for rupture, dissent and the emergence of radical alternatives (Cretney and Bond, 2014; Nelson, 2014). Thus, the enactment of resilience discourse and practice can be heavily interpolated in the multiple and co-existing processes of neoliberal capitalist politics as well as resistance. This chapter explores articulations of resilience that emerge at the everyday scale to challenge a totalising interpretation of the role of the concept in fostering or entrenching political and social relations. By drawing on resilience through the lens of post-capitalist politics the argument is made for the need to privilege the everyday and local scale to understand these radical articulations of transformation and contestation (Bahadur and Tanner, 2014; Nelson, 2014).

To discuss these ideas in more depth examples are drawn on from the frontlines of disaster recovery following the devastating earthquakes that affected the city of Christchurch in Aotearoa New Zealand in 2010/11. The politics of the post-disaster context illuminates these struggles over the meaning and enactment of such a contested concept, particularly as it plays out in response to a significant disruption to the perceived stability of normality. Through drawing on ideas of transformation, the potential for alternative manifestations of resilience are positioned to emerge from the rupture of disaster to create a politics of possibility for life beyond capitalism.

The Entanglement of Crisis Politics and Post Capitalism through Resilience

Resilient individuals, nations and cities are attractive facets of capitalist society. In producing readily adaptable individuals, places, economies and communities, discourses of resilience cultivate traits and approaches that can shift with the demands of market-driven global economy in an increasingly volatile and disaster-prone environment (Joseph, 2013; MacKinnon and Derickson, 2012; Neocleous, 2013). In a world facing uncertainty from climate instability and the increased frequency and severity of disasters, this utopic ideal of ever adaptive settlements, communities and individuals can serve to affirm the norms and values that reproduce and reinforce the status quo. However, these forms of crisis politics and governance through resilience are also entangled with the emergence of new and alternative manifestations of society and economy that can emerge from disaster. This positioning of resilience articulates a need for transformation and change in the face of overwhelming social, political and environmental pressures (Bahadur and Tanner, 2014; Davoudi et al., 2012; Tidball and Krasny, 2014). Thus, while machinic assemblages of resilience produce hegemonic manifestations of neoliberal politics, this multiscalar and multifaceted discourse also produces a contradiction by sowing the seeds of resistance and contestation.

To reinforce and reaffirm the order of society and economy, resilience can act in several ways to internalise and normalise the position of neoliberal capitalism following crisis and disaster. First, by shifting the locus of responsibility for adaptation and survival to the individual and community, resilience discourses articulated through a neoliberal lens allow for the potential for political instability following crisis or disaster to be dispersed (Walker and Cooper, 2011). As the effects of disasters are mediated through the social and political landscape, distancing those in power from the uneven nature of vulnerability and risk becomes an important strategy for safeguarding and responding to threats to political legitimacy (Pelling and Dill, 2006; Tierney, 2008). As Pelling and Dill (2006) note, post-disaster politics is rife with examples of regressive policies and encroachments on democracy and human rights that reflect a desire to maintain political and elite power in the face of disruption.

Furthermore, resilience is often framed as apolitical, thus contributing to significant concern around the implications for depoliticisation and the attempted erasure of power relations (Cote and Nightingale, 2011; MacKinnon and Derickson, 2012). In the disaster context, this form of resilience strategy acts to undermine the political nature of these events to reduce conflict and contestation around response and recovery. The resulting depoliticisation of power and politics in influencing and shaping these processes of risk, vulnerability and uncertainty mould one form of resilience as a machine of neoliberal governance that presents a broad agreement on the solutions to the challenges of crisis governance.

The increasing use of resilience in this context can serve to de-emphasise the role of power in shaping the conditions and effects of disaster while also priming communities for exceptional political responses that consolidate power through neoliberal governmentality. These co-opted forms of resilience seek to cultivate subjects that are governed at a distance to survive, thrive and adapt in a neoliberalised, capitalist social system (Joseph, 2013). This use of resilience as a tool of governance ideologically primes communities and individuals to be self-responsive and adaptive in an increasingly disaster-prone world. This form of governance is aligned with a shift towards an ontology of uncertainty and unknowability which reduces the culpability of those in power to change the underlying factors that influence the effects of disaster (Grove, 2013; Joseph, 2013).

Thus, as a potential manifestation of post-politics, this articulation of resilience crafts an appearance of consensus while extending the hegemonic values of neoliberalism into an increasingly diverse and pervasive discourse of everyday life and politics (Allmendinger and Haughton, 2012; Swyngedouw, 2010). Resilience through disaster response and recovery may be engaged to entrench the normalised state of capitalist society and the associated techno-managerial solutions to crises while also contributing to a pattern of exceptionality where disasters are considered as 'unexpected' or 'unprecedented', and thus justified of exceptional responses (Hewitt, 1983; Swyngedouw, 2010, 2013).

Despite the ostensibly dire context in which resilience is practised in this manner, many of those attempting to experiment and create alternatives to capitalist forms of society draw on and explore the usefulness of resilience in their work. With the opportunity for regressive political change following a disaster, space also emerges for resistance to work through and strengthen pre-existing forms of post-capitalist politics to build new forms of relating to each other and everyday practices (Cretney, 2017; Greenberg, 2014; Head, 2016). While the resilience machine reinforces the status quo and the structures of neoliberal capitalism, there also exists the possibility within this assemblage for rupture, dissent and the emergence of radical alternatives. The critiques of resilience are damning and indeed well deserved, however, it is also integral to understand the multiplicity of interpretations of resilience, particularly those more inclined towards a radical post-capitalist politics (see Bahadur and Tanner, 2014; Cretney and Bond, 2014; Nelson, 2014). These alternative articulations represent the incomplete foreclosure of politics through tactics of depoliticisation and add a cautionary note to critiques which may inadvertently reinforce the hegemony of variants of capitalism (Gibson-Graham, 2006; Larner, 2014).

Theoretically, one of the strongest foundations for more radical articulations within resilience theory is the idea of transformation. Walker and Salt (2012) describe transformation as the capacity to transition or change system states when it is deemed necessary or desirable. The potential for transformation within resilience frameworks remains a potent subject of discussion and debate, as the framework still carries serious normative concerns (Engle, 2011; Hudson, 2010). This is particularly important as it is the dominant values and norms of society and those in power that define what is broadly considered desirable or necessary, potentially limiting the breadth and possibilities for actual change and transformation (Davoudi et al., 2012; Walker and Salt, 2012).

Despite these challenges the concept still holds promise for framing a theorisation of alternative resilience articulations. Nelson (2014, page 6) describes this engagement with transformation as an 'ontology of potentiality' that emerges to encourage and create new ways of being and thinking in society. She notes that the alignment of resilience with neoliberal ideology is contingent rather than necessary, with the concept providing a foundation for more radical and subversive forms of politics to emerge. Innovation, learning and experimenting with these aspects of society are integral to fostering a post-capitalist society. Grove (2014, page 14) also describes this possibility as the potential for 'subversive forms of resilience that use the categories, rationalities, and techniques of resilience to advance alternative forms of life to the responsible, empowered, resilient subjects of neoliberal development'.

This radical potential can be harnessed in different forms of action, particularly those centred on cultivating practices, values and relationships beyond capitalism. Gibson-Graham (2006) argues that by overtheorising neoliberal capitalism, progressive scholars can contribute to the hegemony of these discourses, thus intellectually obscuring alternative configurations of society and economy that can and do already exist (Gibson-Graham, 2006). Through challenging the conflation of the economy with capitalism, this theoretical positioning sheds light on the already existing alternatives that are present within and beyond capitalism at the everyday and local scale (Wright, 2010). These perspectives highlight the importance of a 'politics of possibility' and a 'spirit of hopefulness' towards enacting and facilitating alternative forms of society (Gibson-Graham, 2006, pages xvii–1). By extension, this position allows for the possibility that alternative articulations of resilience co-exist within and beyond regressive and neoliberal manifestations of the resilience machine.

Furthermore, there is possibility in the post-disaster context to influence and instigate positive social change through the space generated by a rupture in perceptions of normality and everyday life. Samuel Prince (1920, page 20) captures this in one of the first accounts of the role of crisis as an agent of change, describing life following disaster as 'like molten metal' in that it facilitates a shift in social life and structure. The potential arising from this lies in a period of intense change and flux in which new values and ways of being in society can be nurtured. In this way the disaster is not only a material event but a 'multiplicity of interwoven, often conflicting, social constructions' (Aradau and van Munster, 2011, page 24). From the assertion that disaster is a time of rupture in the perceived normality of everyday life, a sole focus on resilience discourses that articulate and entrench discourses of neoliberal capitalism may inadvertently obscure different articulations that explore and experiment with potentially radical ways of being in society.

Exploring Multiple Articulations of Resilience after Disaster

As has been argued, the enactment of resilience discourse and practice following disaster is heavily imbricated in the articulation of co-existing processes of neoliberal capitalist politics and resistance. These seemingly contradictory aspects of the resilience machine emerge within a complex context of the historical, economic and political facets that shape the specific manifestations of 'resilience' in any given place. The response to the Canterbury earthquakes in Aotearoa New Zealand provides a snapshot in time and space of these dynamics as they emerged following a devastating series of earthquakes in 2010 and 2011. The most devastating and fatal earthquake in February 2011 in the city of Christchurch resulted in the loss of 185 lives, destroyed a large number of buildings in the central city and caused widespread land damage to residential properties. The city has experienced over 12,000 tremors as a result of the initial earthquake as well as ongoing social and political ramifications surrounding the political approach to response and recovery. In particular, the approach of the government to centralise recovery tasks through a command and control approach has been critiqued for the lack of community involvement and the overuse of government powers (Hayward and Cretney, 2015). However, the community response to the disaster has also received significant attention due to the ongoing role of community-led activities such as transitional architecture, community currencies and urban agriculture in revitalising and reinvigorating the recovery effort (Bennett et al., 2014).

Depoliticising and Consolidating Resilience

In Christchurch, resilience discourse has been infused throughout multiple layers of the approach to recovery from the earthquakes, including aspects of community, local government and central government actions (Hayward, 2013; Vallance and Love, 2013; Winstanley et al., 2015). Examples of this include the creation of a community resilience team that was established in the Canterbury Earthquake Recovery Authority, the city's selection as part of the Rockefeller 100 Resilient Cities project, and the international attention mayor Lianne Dalziel has received for her seminars on the importance of community resilience. The situation in the city has also been complicated by government legislation that introduced wide-ranging powers that allowed central government ministers to intervene in local government politics. This has resulted in a government-led recovery following a top-down command and control approach focused heavily on centralisation and enabled through unprecedented legislation.[1] This context is characterised by the contradictory yet coherent actions of a neoliberal government that acts to both extend and reduce the role of the state in different areas while broadly pursuing values of individualism, the supremacy of the market and the virtues of choice (Brenner and Theodore, 2002).

Consequently, the strategy of the government in Christchurch mobilised a series of articulations of resilience that have framed the way the response and recovery are mediated and justified. In particular, the use of resilience and a 'bounce-back' mentality has supported a wider suspension and truncation of democratic rights and processes in the city. This has included the creation of the earthquake recovery authority as a centralised department that has marginalised local authorities, reduced local input and seen an unprecedented transfer of parliamentary power to the executive branch (Geddis, 2010; Hayward and Cretney, 2015; Hayward, 2013). Hayward (2013, page 4) describes the use of resilience as intimately linked to the ideological drivers of this form of disaster recovery. She notes that these actions are related to the desire to: 'justify authorities making decisions quickly and measuring their impact on recovery by the speed with which the city returns to a "new normal" or experiences "certainty" … In reality, this political speed comes at a steep democratic price.'

While the actions of the government appear, on the surface, to contradict the edicts of neoliberal governance and the reduced role of the state, in fact these policies represent the use of disaster and resilience discourse to foster and encourage a rapid return to the political and economic 'status quo'.

These aspects of government-articulated resilience also affected a shift in how people positioned themselves within the recovery. In particular, this can lead to the valorisation of resilient individuals who are capable, self-sustaining and coping with the emotional and mental upheaval of the disaster experience (O'Malley, 2010). Ideologically, this approach also displaces conflict and contestation over the risk and vulnerability mediated by the disaster and the political and economic decisions made in the recovery.

Some residents of Christchurch felt that the use of resilience by the government to cultivate the idea of a self-sustaining and resilient citizen came at the expense of people's ability to admit they were not coping, or that they were not happy with some aspects of the recovery and rebuild. For example, one resident noted the inability to express discontent in this political environment:

It's almost like there's an expectation that we're all so resilient. What do you do when you [don't] feel like that? Who do you talk to? Does it become not okay to be not resilient? To be really, really, really sick of living in this broken city and have had enough of it. (JA)

Residents also described how this environment appeared to be designed to calm fears of an out-of-control populous as one person described: 'I always got the sense that there was this big fear among the officials … It just felt quite manipulative' (SH). Here, discourses of resilience can be seen to depoliticise the potential for conflict around the shape and form of the recovery, affecting not only the priorities of the rebuild but also the way residents positioned themselves in debates about the future of their city.

It is precisely this depoliticised agenda that represents the continued expansion of neoliberal strategies of governance through resilience discourse. This reinforces and supports the numerous accounts of a resilience agenda driven and strengthened by neoliberal ideology. Furthermore, the agenda engages the techniques and tactics of depoliticisation to craft an appearance of consensus and agreement on the normative value of resilience in the post-disaster context. The post-political governmentality that is fostered through these interventions aims to displace dissent and contestation in the democratic process (Swyngedouw, 2010).

While such a perspective does not deny the presence of politics, which is in fact heightened during times of crisis, it instead points to an attempt to craft the *appearance* of depoliticisation by the political and economic elite to minimise contestation and dissent. Thus, while the use of resilience in this manner justified a centralised response to the earthquakes, something that appears to sit uneasily with orthodox neoliberal thought, there are still clear discursive patterns which are indicative of a neoliberalised governance agenda driven, in part, through an engagement with a depoliticised discourse of resilience.

Recreating and Transforming Resilience

Despite this compelling case, the entrenchment of the economic and political status quo through articulations of resilience that depoliticise and exceptionalise disaster represents only one possibility. Building on the foundation of transformation and an 'ontology of potentiality', the resilience machine also contributes to the emergence of possibility for radical shifts in what is possible in society (Nelson, 2014, page 6). This radical potential for challenging the underlying determinants of risk and vulnerability represents a hopeful engagement with the politics of change, an experiment in what is possible in the cracks of depoliticising strategies and the hegemony of neoliberal capitalism (Head, 2016; Larner, 2014).

Paying attention to alternative configurations of society and economy that emerged within and beyond capitalism in the recovery of Christchurch provides a unique opportunity to render visible the possibility and potential for societal change through alternative forms of resilience. Of particular relevance is the manner in which resilience discourses inform and motivate change that challenges the enclosure of what is politically and economically possible while also maximising the potential for change that emerges following disaster (Cretney and Bond, 2014; Gibson-Graham, 2006).

In Christchurch, community organisations involved in the earthquake recovery at the local scale incorporated resilience into their actions to frame transformative and potentially radical action. This engagement with the politics of locally driven earthquake recovery contested the depoliticised agenda crafted by centralised forms of recovery and repoliticised the foreclosure of participation while also providing entry points to transformative action. As Larner (2014) notes, presenting arguments for the post-political perspective can have the unintended effect of foreclosing the creation of alternatives by articulating a very narrow space for what is considered 'proper politics'. By paying attention to already existing forms of resistance and re-creation that are being enacted more broadly, including through the frame of resilience, the potential for positive change becomes visible.

These forms of community-led recovery were demonstrated in the medium term following the Christchurch earthquakes. Notably, individuals and groups mobilised to establish a network of recovery-based organisations that tackled a variety of issues in the city. This included the emergence of a Transitional Architecture movement that consisted of organisations such as Gap Filler and Greening the Rubble that aimed to install experimental and participatory projects in public and private space left vacant by demolition as well as a Food Resilience Network, which included a variety of suburban and inner-city organisations aiming to utilise the time of recovery to initiative and sustain urban food production and distribution. Other organisations such as Brave New City were established to provide a voice for residents in articulating a vision for the future of the city independent of government-led consultation processes.

Resilience was most notably engaged through neighbourhood residents' organisations and food-based projects. Other organisations did not engage extensively with the philosophy or language of resilience but still contributed to radical and transformative action to create post-capitalist futures in the present. Of the organisations that engaged resilience explicitly, one of the main themes was the use of the rupture of disaster to extend and experiment with the possibilities for change. Here, the emergence of food-based recovery projects has been one of the more visible manifestations of this form of community-led resilience. The processes of gardening, engaging in green space and the natural environment at times of crisis and disaster are increasingly seen as interlinked and interdependent (Okvat and Zautra, 2014; Tidball, 2014).

In Christchurch, these actions were a direct attempt to engage transformational capacity to shift practices and values away from the business-as-usual model engaged in the wider recovery. As several authors involved in an urban agriculture project described: 'This disconnection with the land, and subsequently with our food production, goes back to the founding of Christchurch ... The Central City Recovery Plan and Land Use Recovery Plan are merely reinforcing this same incongruence' (Peryman et al., 2014, page 468).

Tidball (2014) suggests that the interconnection, co-evolution and affiliation of humans with the environment is a strength which can be drawn on to facilitate transformative resilience in the face of trauma and crisis. In Christchurch, this foundation was drawn on to repoliticise the processes and outcomes of recovery, particularly in relation to the environment and food production.

Through this focus on resilience as a form of transformation, communities have been able to look beyond the earthquakes to envision alternative futures beyond the capitalist present. As one individual described 'in terms of resilience ... that's one of the reason's we're focused on the food forests. We're looking for a positive future. We're looking ahead' (BC). Thus, resilience through transformative agricultural and greening projects also articulated a form of resilience that privileged radical change and self-organisation as a response to the destabilisation, destruction and reorganisation that is associated with crisis and disaster (Tidball and Krasny, 2014).

More widely, forms of community-led resilience contributed to shifts in values and norms towards a post-capitalist politics of recovery and reconstruction. In response to the earthquakes, one neighbourhood-based organisation, Project Lyttelton, extended community economy projects that fostered different norms and values while also contributing to material change (Cretney and Bond, 2014). The actions of the organisation through a pre-existing timebank[2] nurtured values of social support and community that stood in opposition to the individualist and self-responsibilisation discourses of hegemonic neoliberal ideology. As an alternative community currency, the timebank also fostered different values and norms around 'work' and the role of money in society, highlighting the role of power in shaping the outcomes of disaster response and recovery.

Importantly, these types of resilience projects shift norms and values in society away from an individualised ideology towards a local community network based around encouraging social justice, support and connection (Cretney and Bond, 2014). Nelson (2014) likens these forms of radical transformation through resilience to resistance, noting that, if the belief in alternatives exists, then those with the belief will affect the creation of new possibilities. While established at a small scale, transformations in these systems that structure capitalism create and contribute to wider alternative systems and structures that grow to larger scales (Nelson, 2014). As authors such as Hosking and Palomino-Schalscha (2016) argue, these practices are often engaged in a relational manner that goes beyond the local. They argue that these 'more-than-local' relations are integral to the transformative potential of post-capitalist action (Hosking and Palomino-Schalscha, 2016, page 5).

In Christchurch, the consequences of these projects have begun to have wider effects on local governance to facilitate ongoing cooperative engagement in formal political processes. Most notably the work of community organisations has led to the city council creating and enacting a food resilience policy and the creation of an urban farm and orchard, part-funded by different local and central government departments, but run by local community groups such as the Food Resilience Network. This has demonstrated that radical articulations of resilience in the city may be expanded in a way that works simultaneously within and outside of formal processes of politics and governance. While there is always the risk of co-option, many of these projects were able to engage with local government in a manner that increased resourcing and support while maintaining autonomy. Importantly, with adequate resourcing and support, community-led resilience projects may be less likely to morph into co-opted forms of neoliberalised resilience that encourages a devolved 'responsibility without power' (Peck and Tickell, 2002, page 386).

Through cultivating a politics of possibility, the centralised articulations of government-led resilience were largely rejected by those at the frontlines of community-led disaster recovery in Christchurch. More widely, resilience was engaged at the grassroots level as a strategy to inspire and catalyse a very different form of disaster recovery that cultivated more transformative shifts in the economy and society. The actions of organisations through projects such as these may act in a way that fosters small-scale shifts in how people see what is possible (Gibson-Graham, 2006). Here, the radical potential of resilience is through actions that highlight underlying power relations in society to repoliticise and focus on practical and constructive ways of living in a different way as individuals and as communities.

Conclusion

This chapter began with an emphasis on the importance of understanding the hegemonic and radical potential of resilience discourse. In the discussion that has followed, these multiple articulations of resilience have been framed through the context of disaster through the case study of the city of Christchurch following a series of devastating earthquakes. As an extension of neoliberal post-disaster politics, the machinic assemblage of resilience can engage with strategies of depoliticisation as part of a wider mobilisation of neoliberal governance (Cote and Nightingale, 2011; Joseph, 2013; Walker and Cooper, 2011). Of concern in this context is the potential for these strategies to inform and facilitate exceptional responses to disaster to consolidate power, particularly in a manner that suspends aspects of democracy and participation. Through the politics of disaster recovery following the Christchurch earthquakes these dynamics have been expressed through government resilience rhetoric that has attempted to cultivate an acceptance for the suspension of democracy through a desire to maintain the economic and political status quo.

However, the presence of these discourses throughout government-led recovery efforts did not fully foreclose contestation surrounding the interpretation and enactment of resilience. As Larner (2014) notes, the post-political perspective can have the unintended effect of foreclosing the creation of alternative forms of politics by articulating a very narrow space for what is considered 'proper politics'. As the engagements with resilience by community organisations demonstrate, it is also possible to see the emergence of more radically oriented grassroots resilience that challenges and destabilises neoliberal and depoliticising narratives. Thus, the resilience machine can produce both the means to extend and entrench normative resilience as well as the potential for the evolution of radical and transformative articulations.

Through a theoretical framing of radical, everyday politics within and beyond the capitalist present, these alternative manifestations of resilience can be understood as part of a wider exploration of already existing alternatives to capitalism. Forms of autonomous community action can embody the experimental and messy process of creating new ways of being and living, including through initiatives such as strengthening community economies, facilitating new systems for food production and distribution and influencing wider processes of local governance (Pickerill and Chatterton, 2006). Discussions around the potential of radical or community-led resilience demonstrate the multiplicity of resilience discourses and strengthen the need to see beyond a singular manifestation of the concept. In doing so, the importance of understanding the politics that informs the radical and regressive interpretations of resilience build on our understanding of the potential for grassroots action to create and resist, particularly following disaster and crisis.

Notes

1 The CERR 2010 Act and the CER 2011 Act were introduced by the government to approach the large task of responding and recovering from the earthquakes. However, the content of these laws has been criticised for the large transfer of parliamentary power to the executive branch, the use of urgency to bypass democratic processes, the potentially unconstitutional nature of the laws and the precedent set by engaging such wide-ranging legislation following a disaster (Geddis, 2010; Hayward and Cretney, 2015).
2 A timebank is an alternative community currency based on the exchange of time credits valued equally for all types of work (Cahn, 2004).

References

Adger, W.N. (2000). 'Social and ecological resilience: are they related?' *Progress in Human Geography*, 24(3): 347–364. Available at: https://doi.org/10.1191/030913200701540465

Allmendinger, P. and Haughton, G. (2012). 'Post-political spatial planning in England: a crisis of consensus?' *Transactions of the Institute of British Geographers*, 37(1): 89–103.

Aradau, C. and van Munster, R. (2011). *Politics of catastrophe genealogies of the unknown*. Oxford: Routledge.

Bahadur, A. and Tanner, T. (2014). 'Transformational resilience thinking: putting people, power and politics at the heart of urban climate resilience'. *Environment and Urbanization*, 26(1): 200–214.

Bennett, B., Dann, J., Johnson, E. and Reynolds, R. (eds) (2014). *Once in a lifetime: city-building after disaster in Christchurch* (1st edn). Christchurch, New Zealand: Freerange Press.

Brenner, N. and Theodore, N. (2002). 'Cities and the geographies of "actually existing neoliberalism"'. *Antipode*, 34(3): 349–379. Available at: https://doi.org/10.1111/1467-8330.00246

Cahn, E. (2004). *No more throw away people: the co-production imperative* (2nd edn). Washington, DC: Essential Books.

Cote, M. and Nightingale, A.J. (2011). 'Resilience thinking meets social theory: situating social change in socio-ecological systems (SES) research'. *Progress in Human Geography*, 36(4), 475–489. Available at: https://doi.org/10.1177/0309132511425708

Cretney, R. (2017). 'Towards a critical geography of disaster recovery politics: perspectives on crisis and hope'. *Geography Compass*, 11(1), e12302. Available at: https://doi.org/10.1111/gec3.12302

Cretney, R. and Bond, S. (2014). 'Bouncing back" to capitalism? Grass-roots autonomous activism in shaping discourses of resilience and transformation following disaster'. *Resilience*, 2(1), 18–31. Available at: https://doi.org/10.1080/21693293.2013.872449

Davoudi, S., Shaw, K., Haider, J., Quinlan, A., Peterson, G., Wilkinson, C., Harmut, F., McEvoy, D. and Porter, L. (2012). 'Resilience: a bridging concept or a dead end?'; '"Reframing" resilience: challenges for planning theory and practice interacting traps: resilience assessment of a pasture management system in northern Afghanistan'; 'Urban resilience: what does it mean in planning practice?'; 'Resilience as a useful concept for climate change adaptation?'; 'The politics of resilience for planning: a cautionary note'. *Planning Theory and Practice*, 13(2), 299–333. Available at: https://doi.org/10.1080/14649357.2012.677124

Engle, N. (2011). 'Adaptive capacity and its assessment'. *Global Environmental Change*, 21(2): 647–656.

Geddis, A. (2010). 'An open letter to New Zealand's people and their parliament'. *Pundit*. Available at: http://pundit.co.nz/content/an-open-letter-to-new-zealands-people-and-their-parliament

Gibson-Graham, J.K. (2006). *A postcapitalist politics*. Minneapolis, MN: University of Minnesota Press.

Greenberg, M. (2014). 'The disaster inside the disaster: Hurricane Sandy and post-crisis redevelopment'. *New Labor Forum*, 23(1): 44–52. Available at: https://doi.org/10.1177/1095796013513239

Grove, K. (2013). 'Hidden transcripts of resilience: power and politics in Jamaican disaster management'. *Resilience*: 1–17. Available at: https://doi.org/10.1080/21693293.2013.825463

Grove, K. (2014). 'Agency, affect, and the immunological politics of disaster resilience'. *Environment and Planning D: Society and Space*, 32(2), 240–256. Available at: https://doi.org/10.1068/d4813

Gunderson, L. (2010). 'Ecological and human community resilience in response to natural disasters'. *Ecology and Society*, 15(2): 1–11.

Hayward, B. (2013). 'Rethinking resilience: reflections on the earthquakes in Christchurch, New Zealand, 2010 and 2011'. *Ecology and Society*, 18(4). Available at: https://doi.org/10.5751/ES-05947-180437

Hayward, B. and Cretney, R. (2015). 'Governing through Disaster', in J. Hayward (ed.), *New Zealand government and politics*. Oxford: Oxford University Press, pp. 403–415.

Head, L. (2016). *Hope and grief in the Anthropocene: reconceptualising human-nature relations*. London: Taylor and Francis. Available at: www.myilibrary.com?id=898284

Hewitt, K. (1983). 'The idea of calamity in a technocratic age', in K. Hewitt (ed.), *Interpretations of calamity* (Vol. 1). Boston, MA: Allen and Unwin, pp. 3–30.

Hosking, E.N. and Palomino-Schalscha, M. (2016). 'Of gardens, hopes, and spirits: unravelling (extra)ordinary community economic arrangements as sites of transformation in Cape Town, South Africa'. *Antipode*. Available at: https://doi.org/10.1111/anti.12259

Hudson, R. (2010). 'Resilient regions in an uncertain world: wishful thinking or a practical reality?' *Cambridge Journal of Regions, Economy and Society*, 3(1), 11–25. Available at: https://doi.org/10.1093/cjres/rsp026

Joseph, J. (2013). 'Resilience as embedded neoliberalism: a governmentality approach'. *Resilience*, 1(1): 38–52.

Larner, W. (2014). 'The limits of post politics: rethinking radical social enterprise', in J. Wilson and E. Swyngedouw (eds), *The post-political and its discontents: spaces of depoliticisation, spectres of radical politics*. Edinburgh: Edinburgh University Press, pp. 189–207.

MacKinnon, D. and Derickson, K.D. (2012). 'From resilience to resourcefulness: a critique of resilience policy and activism'. *Progress in Human Geography*, 37(2): 253–270. Available at: https://doi.org/10.1177/0309132512454775

Nelson, S.H. (2014). 'Resilience and the neoliberal counter-revolution: from ecologies of control to production of the common'. *Resilience*, 2(1): 1–17. Available at: https://doi.org/10.1080/21693293.2014.872456

Neocleous, M. (2013). 'Resisting resilience'. *Radical Philosophy*, 178(6): 2–7.

Okvat, H. and Zautra, A. (2014). 'Sowing seeds of resilience in community gardening in a post disaster context', in K. Tidball and M. Krasny (eds), *Greening the red zone: disaster, resilience and community greening*. Dordrecht: Springer Publishers, pp. 73–90.

O'Malley, P. (2010). 'Resilient subjects: uncertainty, warfare and liberalism'. *Economy and Society*, 39(4): 488–509. Available at: https://doi.org/10.1080/03085147.2010.510681

Peck, J. and Tickell, A. (2002). 'Neoliberalizing space'. *Antipode*, 34(3): 380–404.

Pelling, M. and Dill, K. (2006). '"Natural" disasters as catalysts of political action'. *Human Security and Resilience, ISP/NC Briefing Paper*: 4–6.

Peryman, B., Peryman, O. and Marquet, M. (2014). 'Digging where we stand: visions of urbundance and the role of food production in Christchurch', in B. Bennett, J. Dann, E. Johnson and R. Reynolds (eds), *Once in a lifetime: city-building after disaster in Christchurch* (1st edn). Christchurch, New Zealand: Freerange Press, pp. 467–470.

Pickerill, J. and Chatterton, P. (2006). 'Notes towards autonomous geographies: creation, resistance and self-management as survival tactics'. *Progress in Human Geography*, 30(6): 730–746.

Prince, S.H. (1920). *Catastrophe and social change, based upon a sociological study of the Halifax disaster*. New York: Columbia University.

Swyngedouw, E. (2010). 'Apocalypse forever? Post-political populism and the spectre of climate change'. *Theory, Culture and Society*, 27(2–3): 213–232. Available at: https://doi.org/10.1177/0263276409358728

Swyngedouw, E. (2013). 'Apocalypse now! Fear and doomsday pleasures'. *Capitalism Nature Socialism*, 24(1): 9–18. Available at: https://doi.org/10.1080/10455752.2012.759252

Tidball, K. (2014). 'Urgent biophillia: human-nature interactions in red zone recovery and resilience', in K. Tidball and M. Krasny (eds), *Greening the red zone: disaster, resilience and community greening*. Dordrecht: Springer Publishers, pp. 53–71.

Tidball, K. and Krasny, M. (2014). 'Resilience and transformation in the red zone', in K. Tidball and M. Krasny (eds), *Greening the red zone: disaster, resilience and community greening*. Dordrecht: Springer Publishers, pp. 25–44.

Tierney, K. (2008). 'Hurricane Katrina: catastrophic impacts and alarming lessons', in J. Quigley and L. Rosenthal (eds), *Risking house and home: disasters, cities and public policy*. Berkeley, CA: Berkeley Public Policy Press, pp. 119–136.

Vallance, S. and Love, R. (2013). 'The role of communities in post disaster recovery planning: a Diamond Harbour case study'. *Lincoln Planning Review*, 5(1–2): 3–9.

Walker, B. and Salt, D. (2012). *Resilience practice: building capacity to absorb disturbance and maintain function*. Washington, DC: Island Press.

Walker, J. and Cooper, M. (2011). 'Genealogies of resilience: from systems ecology to the political economy of crisis adaptation'. *Security Dialogue*, 14(2): 143–160.

Winstanley, A., Hepi, M. and Wood, D. (2015). 'Resilience? Contested meanings and experiences in post-disaster Christchurch, New Zealand'. *Kotuitui: New Zealand Journal of Social Sciences Online*, 10(2): 126–134.

Wright, S. (2010). 'Cultivating beyond-capitalist economies'. *Economic Geography*, 86(3): 297–318. https://doi.org/10.1111/j.1944-8287.2010.01074.x

6

THE RESONANCE AND POSSIBILITIES OF *COMMUNITY* RESILIENCE

Lauren Rickards, Martin Mulligan and Wendy Steele

Introduction

A vast literature now describes and demonstrates the appeal and challenges of the multifaceted concept of resilience. But why is *community* resilience in particular so popular – and contested? While often paired with resilience as a mere scalar or spatial unit synonymous with the local, 'community' is a deceptively complex term with a long history of debate. In this chapter, we explore the complexity and progressive unbounding of the concept of community in order to highlight its diagrammatic similarity to and intersections with similar efforts to unbound the concept of resilience.

Resilience Meets Community

With the 'resilience machine' now in full swing and resilience initiatives popping up around the world, it is widely evident that not only is 'actually existing resilience' far from a 'universal resilience project' (Grove and Adey, 2015, page 78), but it is highly contested. To date, the concept of resilience has faced two broad lines of critique, one which takes aim at incoherence and imprecision in its use, and one which challenges its presumptions about how the world works. In this introduction we outline these lines of critique in order to identify the diagrammatic resonance and tension between community and resilience and begin to think through what their pairing in 'community resilience' means.

For critics frustrated with not just the ubiquity but the incoherence and imprecision of 'resilience talk', there is now a tendency to demand more information about what exactly proponents have in mind when they call for resilience. It is common, for example, to demand to know 'resilience to what?' and, more significantly, 'resilience of what or whom?' (e.g. Cutter, 2016). While some of this is about important context-specific debates about which exact groups or things are privileged by a specific resilience initiative, at a more general level a common shortlist of subjects and objects tends to be paired with resilience, akin to design choices that the 'resilience machine' can be programmed to produce. These options are now familiar terms: *urban* resilience, *city* resilience, *rural* resilience, *social* resilience, *organisational* resilience, *individual* resilience, etc. Thus, although resilience thinking has helped bring to the fore some relatively new objects such as 'critical infrastructure', it has also reinforced the apparent salience of many of the conventional social and spatial units by which we know the world.

Among the existing lenses reinforced by resilience thinking, one of the most popular and oldest is 'community'. As a unit that can be read as a social or spatial entity, community is an unsurprising partner for resilience. More than just an instrumental fit, community has a certain normative and conceptual resonance with resilience. The idea that community should be supported by resilience efforts, and that resilience should be the goal of community, has, in many parts of the world at least, a common-sense feel.

To better understand this affinity, we follow Grove and Adey (2015) in turning to the aesthetics of resilience. By 'aesthetics' Grove and Adey are referring to Jacques Rancière's notion of aesthetics as 'a delineation of spaces and times, of the visible and invisible, of speech and noise, that simultaneously determines the place and the stakes of politics as a form of experience' (Rancière, 2004, page 8). Resilience, they point out, does aesthetic work. It creates a 'sensorial regime' (Grove and Adey, 2015, page 80), such that 'singular events', whether a fire, flood, food scare, financial crisis or outbreak of fighting, can be 'slotted into specific orderings of time and space that make "sense" through the language, imagery and codings of resilience thinking' (page 79). More than just shaping how we make sense of particular situations, resilience thinking has a topological, diagrammatic quality. It provides a map of abstract reference points, 'sets of relations between relations', to encourage us to orient ourselves in the world in a particular manner. Such diagrams, Grove and Adey explain, drawing on Guattari (1995), are 'self-organising, autopoietic machines that produce specific kinds of worlds, and subjectivities that inhabit those worlds, by articulating material and enunciative contents together in specific ways' (page 81). The 'resilience machine' can be thought of as such a machine, encouraging varied but generally complex systems-based maps of the world, supported by the related concept of adaptation which is similarly promoted as a new guide for living in a climate-changing world (Watts, 2015).

On their own, diagrams are unproductive. To produce their programmed worlds, they need to plug into the existing 'material and enunciative contents' of the world. As Grove and Adey put it, as a diagram 'resilience has no constitutive power of its own: it produces its diagrammatic effects only to the extent that it is able to appropriate affective relations and direct their force towards the production' of a certain type of world (Grove and Adey, 2015, page 82).

We come then to the role of community. As a specific target and partner for resilience – one that is simultaneously spatial and social – the concept of community provides a ready mix of 'material and enunciative contents' for the resilience diagram. Moreover, it has aesthetic and diagrammatic qualities of its own that both resonate with and complement the dominant complex systems imaginary of resilience. As a 'warmly persuasive' concept (Williams, 1976, page 76), community reinforces the naturalness and normative appeal of resilience, pushing its cybernetic tendencies towards the biological and organic rather than the machinic and metallic. It offers a more enlivened, collective, caring version of resilience than that which emerges when the resilience diagram is plugged into the militaristic or neoliberal contents of the world such as 'critical infrastructure resilience', or the nihilistic, radically individualistic form of resilience demonstrated by the character Walt Whitman in the TV series *Breaking Bad* (Grayson, 2017). Although community is far from the only version of the world being (re) generated by the resilience machine, its popularity points to its affinity with resilience, giving community resilience a central, 'proper' place within the resilience agenda. Indeed, policy often appears to be 'grasping' at communities (Grove and Adey, 2015, page 80) and even the concept of community in general.

David Chandler's reading of 'resilience thinking' (to be discussed later) suggests that at work here is more than an opportunistic or pragmatic use of community as a useful handle for resilience initiatives. Instead, Chandler reads in resilience a fundamental reliance on community and individuals as the only groups or level of organisation at which resilience is ever possible. Chandler reads resilience as part of a broader realisation of the (growing) complexity of the world and acceptance of the profound limitations on knowledge and governance that this complexity imposes. Thus, it is only at the level of individuals and communities that the remaining option – emergent self-organisation – is genuinely possible. As he puts it, in resilience thinking 'It is alleged that only by returning power to the individuals and communities, who really have the power to self-organise in relation to the problem, can complexity become a force for governance rather than a barrier to it' (Chandler, 2014, pages 38–9).

At the same time, decades of scholarship on community challenge assumptions about the coherence and functionality of communities. In this view, the concept of community and especially real communities in their complexity and particularity will always exceed the resilience diagram's conventional neat systems boundaries. As is discussed later, critiques of community have pushed scholars to conceive of it in more and more open terms, even reformulating it as at base an essential quality of human existence itself.

When community is reimagined as profoundly unbounded, its diagrammatic alignment with resilience seems to be unsettled. But as the reference above to what Chandler calls 'resilience thinking' and a broader embrace of complexity indicates – and as scholars such as Simon and Randalls (2016), Reid and Botterill (2013) and Anderson (2015) have argued – resilience is multiple. And one way in which it is multiple is in the dimensions of its diagrammatic form; that is, in how its sets of relations are structured. Increasingly, ideas about resilience themselves fit awkwardly with the dominant resilience model of 'bounce-back-ability'. A key reason for this is that the tight coherence of the implicit system that is imagined as having some degree of resilience has been progressively challenged. In line with the critiques of community, its boundaries have been pried open, drawing upon lesser-known aspects of the genealogy of resilience. In this way the diagramming of resilience has shifted towards a less closed, neat and relatively rigid imagined system towards one that is more open, messy and liquid. The resultant new flexibility of resilience appears '"new age" in the face of hard, macho systems'(Grove and Adey, 2015, page 80). This loosening of system boundaries towards a more fluid, 'contemporary' set of relations represents a kind of reprogramming of the resilience machine, a reconfiguration of the resilience diagram along more malleable, unpredictable and evolutionary lines, opening up the possibility for – though not a guarantee of – more inclusive and transformative outcomes.

One of the striking things about this reprogramming of the resilience machine is that in many ways it intensifies, not diminishes, the resonance between resilience and community aesthetics. Although the literatures about the unbounding of each concept have – with the partial exception of David Chandler (2014) – barely been brought into contact, they are highly synergistic in many ways. The redrawing of resilience in terms of more open, ambiguous, evolving systems helps counter the tension that otherwise exists in the notion of community resilience if community is conceived as an open-ended entity. It enables the community and resilience diagrams to once again better align. The existence of these tensions and alignments rests on an underlying sense that both community and resilience are diagrammed along similar spectrums from more to less bounded forms (see Figure 6.1).

We do not want to imply with the above that there is a simple synergy between different types of community or resilience aligned at similar points along the spectrum. Virtual communities, for example (those sets of relations that bring distant digital users into a sense of community), do not *necessarily* respond to disruptions in terms of its corresponding term in Figure 6.1: 'evolutionary resilience' (Davoudi et al., 2013) (a resilience process in which new forms emerge following disruption, and the latter is

FIGURE 6.1 The closed-open spectrum common to the diagrammatic form of community and resilience
Source: Authors' own

conceived as normal not exceptional). Rather, we offer the above in a speculative mode and suggest that appreciating the similarities in the diagrammatic form of community and resilience – that is, the way both can be conceived along a bounded-unbounded axis – is a useful heuristic for exploring what is implied in their pairing as 'community resilience'. It allows us, for example, to identify four possible intersections (different combinations of bounded versus unbounded versions of resilience and community), which potentially illuminate some of the different forms of community resilience that exist and the aesthetic work they do (see Figure 6.2).

To what extent the pairing of community enhances/exacerbates the unbounding of resilience and vice versa is yet to be seen. Does the concept of community provide an anchor for resilience within an otherwise tumultuous world? Does the concept of resilience provide some much needed security to individuals in radically unbounded communities? Or do they interact in different ways entirely? To contribute to this exploration, in this chapter we outline efforts to progressively open up the community diagram, exposing the politics implicit to its dominant and specific forms.

Unbounding Community

Conventionally, 'community' refers to a set of relations that somehow connects a group of people. In their search for a consensus definition to use in public health MacQueen et al. (2001), for example, define community as: 'a group of people with

	Resilience diagram	
Community diagram	**Bounded**	**Unbounded**
Bounded	Cohesive place-based social groups that return to *status quo* if disturbed?	Strong social entity collectively engaged in an ongoing process of transformation?
Unbounded	Individuals engaged in volatile but continuous processes of re-becoming?	Transitory collections of people in the midst of emergence and flux?

FIGURE 6.2 Speculative schematic of bounded and unbounded diagrams of community intersecting with those of resilience

Source: Authors' own

diverse characteristics who are linked by social ties, share common perspectives, and engage in joint action in geographical locations or settings' (page 1929). Consensus about community, however, has been hard to find. Notions of community vary in numerous ways, including whether an external or internal or subjective view of community is adopted, or whether a qualitative, felt, processual sense of community versus a more quantitative, functional, outcome one is inferred. Broadly speaking, at least two major genealogical origins of community can be detected. One is the idea of a pastoral flock and political community encouraged by the church and the state, overlaid with notions of place associated with emerging geographical settlement patterns such as towns and nation. The second is in ecology, where the early 20th-century idea of plant and animal communities as detectable, place-based assemblages of species was, at least originally, tightly entwined with systems thinking and Frederick Clements' organismic notions in particular. Both the pastoral and ecological origins of community evoke an aesthetic of belonging, helping explain why the term community 'never seems to be used unfavourably' (Williams, 1976, page 76). Yet, as pointed out by searing critiques of the assumptions of sweet harmony in religious, political and ecological communities, the notion of community and especially context-specific evocations of community are not as inherently wholesome, innocent or self-evident as many politicians like to pretend (Mulligan et al., 2016). Although politicians often use the word community to suggest that their policies stand above partisan political contestation, many scholars have pointed out that the search for community is always a matter for political contestation (Rose, 1996; Bauman, 2001; Esposito, 2010; Brent, 2009). In the remainder of this section, seven lines of critique are outlined that have been directed at the conventional notion of community as a pre-given, harmonious entity.

The first criticism levelled at the conventional notion of community has been about the exclusivity its normalised boundary poses. Many scholars (e.g. Delanty, 2003) have argued that community only comes into existence when it is consciously constructed and performed. Assumptions about the existence of a self-evident 'community' can elide tricky equity questions about who is recognised and valued and who is not. While the aesthetic of community is often one of warm inclusion, in practice it imposes a diagram on the world in which some people are in and some are out. In some cases, the exclusions are ontological and epistemological, generated by blind spots, imprecision or sloppiness in map making. In other cases, the exclusions enacted are strategic, if not conscious, as illustrated by the persistent marginalisation of women from implicit workplace communities in many industries (e.g. Heggem, 2014). Still others are generated by the deliberate controlling of borders, with violent delineations of membership enacted by intersecting biopolitical and geopolitical processes along spatio-political boundaries that unjustly shut out many (Hyndman, 2012).

Questioning the assumed desirability and not just the accuracy of a particular community is the second move to open up thinking about communities. Although certain groups may be included in a particular representation of community, and such representations reinforce the aesthetic appeal of community, in practice their inclusion may come with various forms of distributive, procedural and cultural injustice as others work to produce or 'protect' a certain form of community. This includes well-meaning efforts to employ more participatory forms of policy making that are often motivated by a desire to improve future outcomes for a community, but that in doing so begin with an assumption that such community exists and is valued by those involved. As Pearson et al. (2016) found in their effort to engage local communities in scenario planning exercises, for example, members of a 'community' can have very different perceptions of the given community's past and present, and thus very different ideas about what a desirable future looks like. Grove (2013) similarly found this in Jamaica: 'Faced with ontologically distinct experiences of vulnerability and insecurity, community members' visions of empowerment and resilience – what they "want to do" – may not align with those of policymakers and fieldworkers' (page 203).

Many studies of post-disaster recovery work have found that well-meaning international agencies often trigger conflict and division within targeted local communities (e.g. Mulligan and Nadarajah, 2012). The lessons have long been raised by those involved in development work. Writing about work to reform land tenure systems in the rural South African context, for example, Kepe (1999) notes that in using the term 'community': 'Effects are positive when they help focus policy on the needs of poor people, but negative when they force conflicting groups together in a manner which results in the rights of a weaker group being trampled on by the actions of a more powerful group' (page 415).

These questions of justice bring us to the third question raised about given renditions of community, which is who has the power to name and describe a community. Many people – certain groups and communities more than others – find themselves subject to representations of their community imposed by more powerful others, such as those with their hands on the levers of resilience policy machines. In such cases, those represented may contest how their community is represented. Drawing on 25 years of community development practice, Jeremy Brent (2009, page 261) notes that while community cannot be 'an answer to oppression' it is commonly 'a form of resistance within asymmetrical relations of power'. Nevertheless, the term 'community' is increasingly used by government and other agencies as a synonym for 'target audience': a group of heterogeneous individuals in whom some kind of (behaviour) change is wanted (e.g. to increase their and the nation's resilience), irrespective of whether felt 'communal' relationships actually exist between them (Pudup, 2008; Ernwein, 2014). Indeed, Esposito (2010) and Jha (2010) argue that clumsy, narrow projections of community identities can exacerbate social division and conflict. An illustration of this is provided by Ichinkhorloo and Yeh (2016), who found in their study of a community-based natural resource management project that 'rather than trust, the formation of these new "communities" creates opportunities for struggles over power, influence and resource access' (pages 3–4). A further illustration of this is provided by Sandhu (2016). Writing from an intersectional feminist perspective, she reflects on an invitation she received in 2014 'to join a Sikh Community Resilience Forum for Coventry ... as a member of the "Sikh community" in the city'. The aim of the forum was for the police: 'to communicate messages to the "Sikh community" in order to achieve an improved way of working together so that communities can be safe and ensure sensitive issues and concerns, which may impact on cohesion in the city, would be raised and discussed' (page 89). Sandhu has 'never identified solely as a Sikh woman' (page 88). She declined the invitation out of a concern that participating would endorse the belief that faith-based 'community leaders speak for the community', collapsing other important axes of inequality such as gender and class 'into one with the label of "faith"' and undermining her efforts to fight the 'erosion of women's spaces and rights' (pages 96 and 98). The fact that it is police initiating this dialogue is of significance, pointing to the sense that functional coexistence requires that people adhere to certain ways of doing things. While indisputable, it is also clear that rules alone are insufficient in creating community. Indeed, they may undermine a more voluntary sense of community of the sort that reduces people's desire to 'do the wrong thing' in the first place. Many critics have pointed out that a law-based enforcement of 'community' can unhelpfully perpetuate divisive identities that undermine rather than foster true community. They have called instead for approaches to community based on the celebration of diversity and difference, spontaneous communities of choice, not imposed communities of necessity.

We come then to the more general criticism of what is meant by the word 'community'. Beyond problems with describing any particular community, this fourth point is about the problem of defining community in general. Assumptions that communities are, by definition, about solidarity and cohesiveness have been dismissed by some as old-fashioned, challenging the warm aesthetic of community and seeking to replace it with something fresher and lighter. That very different people are increasingly being 'thrown together' is a truism of our contemporary globalised existence. Increasing global flows of people, goods, information and ideas have strengthened (and possibly necessitated) resistance to neatly homogenous notions of community (Delanty, 2003; Rose, 1996; Bauman, 2001). Any grouping of people is increasingly likely to include individuals with diverse histories, allegiances, daily rhythms and experiences. Work on transnationalism by scholars such as Doreen Massey (2005) underlines the ongoing creation of new experiences of community at the level of locale. Within wider conversations about place identities – those contextualised and inevitably politicised processes of intersecting local and national identify formation – individuals coexisting in a given locale find themselves encountering many unfamiliar 'others' with whom they are obliged to negotiate forms of coexistence, however transient. This more dynamic experience of community-in-the-making leads Brickell and Datta (2011, page 8) to call for us to 'explode a notion of insularity invested in the locale' and instead to see how in the context of rising personal mobility many contemporary place-based communities are inherently open. Echoing Massey, Brickell and

Datta note that people on the move can carry forms of accumulated social capital which enable them to participate in community making wherever they might be residing, with community making therefore becoming a more 'translocal' enterprise. In this reading, community has become as much a border-crossing activity as a border-maintaining stance, at least for those with the social and financial capital to cross borders, such as the skilled workers Ong (2005) describes as moving easily across national borders while local unskilled workers are denied not just mobility but genuine citizenship in their one place of residence.

The analytical focus on place-based communities is, then, the fifth way in which the idea of community is being challenged. While place-based communities remain relevant to the extent we have homes in one place and feel connected to others in the area, for many people the mobility of modern life (whether for work, leisure or escape) and complexity of modern identities means that other more dispersed social groups may be equally if not more important to our sense of community. As Noel Castree (2004, page 135) has argued, we need to adopt a 'stretched' understanding of locale. Any one place may contain a host of subcommunities that reach well beyond the boundaries of place, forming for example extensive transnational diasporic communities, while also relying as a community upon constant connections to higher-order, distantly located governing bodies.

Some of this rescaling of the community diagram is topological in nature, emphasising the relevance of relations between spatially distantiated places. But some is more conventional, expanding what counts as community to larger spatial and social levels a shared sense of belonging is more important than regular interactions – that is, a common sense of being part of something shared. One common-sense focus for such belonging is the nation, which is the community on which Benedict Anderson famously introduced his notion of the 'imagined community' (Anderson, 2006); one of the theories about community that most points at its aesthetic, diagrammatic quality. Like the community diagram more broadly, an imagined community does not have to plug into the nation, though; it can refer to any imagined group that people identify with, be that other governance units such as catchments that people are encouraged to identify with, or communities based around shared concerns, activities or minority status. People may feel a greater sense of belonging to their professional community, for example, than they do to the loose collection of people who happen to live in their area at any particular moment. Or they may feel a strong sense of belonging to a political community: a group of people connected by a shared desire for certain change in the world. As Sandhu's story above suggests, being cordoned into someone else's imposed, politically motivated, imagined community, such as a nation state or target faith-based population, might, ironically, strengthen a countervailing connection with a resistive political community, pointing to the way that political communities intersect in complex ways with questions of who is and is not recognised (Staeheli, 2008).

Sixth, many imagined communities are underpinned not by a shared (residential) locale as much as engagement-at-a-distance. Many of us now enjoy a plethora of close relations with people far afield, some of whom we have never met in person. New communication technologies have greatly increased our ability to participate in a host of 'virtual' communities operating at scales ranging from the local to the global, some of which may be small and tight (e.g. a support group) and others which may be large and loose (e.g. eBay). Perhaps in response to 'the fragmentation of society' and resultant 'worldwide search for community', rising demand for 'global forms of communication' has 'facilitated the construction of community' in the digital realm (Delanty, 2003, page 193). Thus, while many people struggle to find any community that will embrace them, others have unprecedented choices in regard to the real or virtual communities that they can participate in, stretching in scale from the micro to the global. Such connectedness places us in a close relation with technology as much as with other people, shifting our subjectivity towards Haraway's figure of the cyborg, who she notes 'defines a technological polis' and 'does not dream of community on the model of the organic family' (Haraway, 2006).

While all virtual relations are unavoidably grounded in various places by their reliance on and uneven access to physical infrastructure, resources flows and capital, not to mention real participants, real designers and governance (Zebrowski, 2015; Parikka, 2015), some virtual communities intentionally divert members' attention away from their immediate surrounds, offering escape from 'the real world' more generally. Other virtual communities have more direct and complex relations with the real world. For example, some virtual communities have a celebrated role in fostering the resilience of place-based communities during and after disaster events. Focusing especially on the use of social media, a rapidly growing literature highlights how disrupted place-based communities' efforts to reorganise are assisted by online communications and support, often between groupings of people that emerge spontaneously in response to the event (e.g. Givoni, 2016; Albris, 2017). Merging the interpersonal and the technical, the capacity for such bonding, and especially bridging, social relations to emerge upon demand is conceived in a resilience framework as not just social capital but 'social infrastructure' (Sadri et al., 2018; Aldrich, 2017). Such phrasing points to not only the perceived utilitarian value of such connections but the motif of webs, networks and flows that replaces that of bounded spaces in virtual communities, with networked communities now read as an expression of resilience (Chandler, 2014).

Far-reaching virtual 'communities' of either a transitory or longer-term kind are also forming between some individuals out of a shared commitment to progressive change 'on the ground' among (other) 'real' communities. In some such communities, such as the Transition Network that uses 'open-access wiki websites and blogs' to share ideas drawn from and directed to real place-based transition towns, many members of virtual communities-of-concern may have little actual connection with their places of concern, valuing them instead for their existence value or as a matter of principle. In contrast to virtual communities generated around a specific place-based community or their disaster-specific needs, most activist-based virtual communities such as the Environmental Justice Network are more focused on commonalities across place-based communities and the chronic and systemic stresses they share. In taking this approach, the virtual community broadens its relevance to potential members and acts as a third party forging imagined links between place-based communities that may, at least initially, be in little direct contact themselves. Illustrating such a virtual, political community is the World Social Forum, which 'provides a set of beliefs and practices about justice and injustice that unites global activists into a community affirming a common identity and devotion to a more just world' (Langman, 2005).

This focus on the global brings us to the final line of argument directed at 'community': a profound questioning of the assumption that community is necessarily about any discrete subset of people at all, but is rather a basic condition of all humans. Esposito (2010) argues that the desire to experience community represents a fundamental fear of 'the hole into which the common thing continually risks falling' (page 8), a desire to belong to a safe and functioning whole. Jean Luc Nancy similarly generalises community to human existence, but strongly resists the sense of a functional whole, rejecting anything that 'fosters a sense of closure, continuity, unity and universalism' (Devadas and Mummery, 2007), including the essentialism of Benedict Andersons's imagined national communities. Like Agamben (1993) with his idea of the 'coming community', Nancy aims to instead stage a radical 'opening up of the idea of community', conceiving of a community that emerges out of action and does not pre-exist it. This 'community without community' is not reliant on a stable essence (the second community in the phrase) but is always only becoming; a community that 'ceaselessly works to produce more democratic, open and fluid relationships with others to foster a sense of "being with" not "being in"' (Devadas and Mummery, 2007, page 23). This reorganisation of community fosters what subaltern studies calls 'horizontal affiliations', new associations open to differences and contradictions (Chakrabarty, 2000). In a community conceived as 'being singular plural', we are all continuously co-becoming, at once our individual and collective selves (Nancy, 2000).

Nancy's linkage of community to a global condition, indeed species condition, resonates with moves to interpolate us as members of a 'global community', as part of a single humanity, evident in some responses to the global environmental crisis and the emergence of 'the Anthropocene' in particular. This globalisation of the notion of community and reinsertion of nature actually resonates with Neil Adger's influential idea of social resilience as not only social in the sense of mediated by social relations such as institutions, but social in the more fundamental sense of being about humans and our relationship with the non-human world. Although Adger's focus was on humans' use of 'natural resources' (an inherently social framing of nature), the idea of social resilience in his work and others has evolved to be predominantly focused upon nature's impacts upon us, defining our membership of the human community on the basis of our shared vulnerability to nature's increasingly volatile behaviour. As Grove and Chandler (2017) discuss, it is this perceived volatility of the 'wild world' we now inhabit – one in which the balance between stability and change has tipped strongly towards the latter, making change the norm rather than the exception – that has helped highlight the apparent need for resilience. Or as (Simon and Randalls, 2016) put it:

> The remarkable prevalence of the resilience concept is tightly bound to the adage that we now live in a 'time of crisis'. It has come to stand for the ability to absorb, withstand, persist, and even thrive and reorganise in the face of the shocks and disturbances of always uncertain becoming, that is now even 'more so'.
>
> *(Page 3)*

According to David Chandler, community has a privileged position in this vision of a world in crisis. For, reflecting the sense that not only is there no stability, there is also no 'outside' to the system, reliance on the wise oversight and sober planning by government is naive. At the same time, individualism is problematised because such isolation is considered ontologically impossible and normatively illogical in a world of relations (Chandler and Reid, 2016), while the capacity of markets to tame 'life' is dismissed as quaint and ignorant of people and nature's inherent resilience know-how (Chandler, 2014). Overall, according to Chandler (2014):

> From the perspective of resilience-thinking, the governance of complexity ... needs to reject the artifice of imposing goals and direction on the world and instead seeks to find its goals in the processes, practices and communicative interactions of the world itself. Resilience-thinking seeks to access, tap into and instrumentalise the 'real' power of life as complexity rather than hubristically ignoring the realities of the complex post-modern world or seeing governance as impossible in the face of a turbulent and uncertain world.
>
> *(Page 37)*

Into this governance vacuum steps community. In resilience thinking it is only community, Chandler suggests, that is seen as allowing complexity to 'become a force for governance rather than a barrier to it' (Chandler, 2014, page 39). In particular, empowered, networked, self-organising, often virtual and even more-than-human *communities* are presented as iconic sites for emergent resilience-in-action.

Strikingly, Chandler identifies in this turn to community an uptake of Jean Luc Nancy's notions of community as co-becoming. According to Nancy, real power does not operate in the conventional model of elites wielding power upon the rest of society, because formal politics fails to 'ever capture the reality of community at the real and ontological level'. Rather, real power exists only in 'the real politics of the life world', in the deeper 'le politique' of community life (Chandler, 2014, pages 59–60). For our purposes, what is most significant about this argument is that it points to a rare intersection between the community and resilience literatures, underlining the relevance of the overlaps we propose exist in how community and resilience have been progressively unbounded. We turn now to some reflections on what this means for the more precise concept of 'community resilience'.

Rethinking 'Community Resilience'

In this chapter we have looked at resilience as a diagram that not only imports pre-existing ideas about the existence of community into its machinic assemblage to feed the production of resilience programmes, but one that is paralleled in extensive ways by an equivalent diagram of community. There is a stylised homology between the ways resilience and community are both conceived as neat and bounded in some renditions, and messy and open in others. Genealogical work is needed to identify the origin of these similarities and whether their common basis in ecology and perhaps even in religious thought points to moments of literal co-becoming. At a conceptual level both concepts are arguably reliant upon the other, with community providing content for resilience and resilience providing ontic stability to community.

At the same time, however, both community and resilience exceed the other. Resilience may not only be about numerous other social and spatial units, but the resilience aesthetic can invoke an ideal completely opposed to community, as seen in the atomised subject *Homo resilis* – 'a subject who is defined by privileging survival through associality' – that Grayson (2017, page 31) describes in his analysis of the TV series *Breaking Bad*. Community is similarly about more than resilience. Community may be, for instance, potentially transitory and ephemeral, an event more than a norm, a disturbance more than a response, a momentary encounter more than an endless cycle. Rather than the foreground as in resilience thinking, community may be a dimly discernible knot in a larger, messier intersection of crashing trajectories, one lens on a far more unruly world.

So, what does all this mean for the idea of 'community resilience'. While far from complete, we want to posit four speculative conclusions to end. First, we suggest, community resilience *is* a privileged form of resilience because of the deep diagrammatic resonance that the two concepts have. One of the reasons for the appealing, natural aesthetic of community resilience is that the two terms are in many ways conceptual twins. Second, an important commonality between them is the way that both have proven resilient to efforts to unbound them by unsettling their seemingly foundational organismic base. Both concepts persist despite successive rounds of fierce criticism because they have been able to re-emerge in new, less rigid, more contemporary forms. This shared resilience has further strengthened their magnetic twinning. Third, the concept of 'community resilience' is itself resilient because of the diverse perspectives, preferences and situations it encompasses (Figure 6.2). When combined in a quadrant (Figure 6.2), 'community resilience' takes on very different hues. In one corner, is the familiar ideal of cohesive strong entities that, if ever disturbed, immediately pulse back to the way they were. In the opposite corner is the alternative ideal of unstructured, emergent collections of people together unconsciously responding to and creating multidimensional change. In between are cohesive though perhaps virtual communities co-generating emergent transformations through open systems of exchange and communication of the sort described by Chandler; and imagined forms of ontological co-becoming based on transitory relations that nevertheless collectively provide a relatively stable base for individuals in the midst of a violently fluxing world. How these ideal types align with real-world community resilience initiatives is an important question for empirical research. Finally, much remains to be done to think afresh about the politics of community resilience and to identify neglected possibilities for more progressive versions of community resilience in given situations. As the (community) resilience machine continues to ramp up, perhaps the most important message about either concept is that they are political, in theory and in practice. Such politics is unavoidable and potentially transformative of the broader global world of crisis we are in, even as the crises we may wish to highlight differ from those of formal politics. The point is that as forms of aesthetics that shape what we do and do not sense, community, resilience and community resilience are not innocent but not valueless either.

References

Agamben, G. (1993). *The coming community*. Minneapolis, MN: University of Minnesota Press.
Albris, K. (2017). 'The switchboard mechanism: how social media connected citizens during the 2013 floods in Dresden'. *Journal of Contingencies and Crisis Management*. Available at: https://onlinelibrary.wiley.com/doi/abs/10.1111/1468-5973.12201.
Aldrich, D.P. (2017). 'The importance of social capital in building community resilience', in W. Yan and W. Galloway (eds), *Rethinking resilience, adaptation and transformation in a time of change*. New York City: Springer, pp. 357–364.
Anderson, B. (2006 [1983]). *Imagined communities: reflections on the origin and spread of nationalism*. London: Verso.
Anderson, B. (2015). 'What kind of thing is resilience?' *Politics*, 35(1): 60–66.
Bauman, Z. (2001). *Community: seeking safety in an insecure world*. Cambridge: Polity Press.
Brent, J. (2009). *Searching for community: representation, power and action on an urban housing estate*. Bristol: Policy Press.
Brickell, K. and Datta, A. (2011). 'Introduction: translocal geographies', in K. Brickell and A. Datta (eds), *Translocal geographies: spaces, places, connections*. Burlington, VT: Ashgate, pp. 3–22.
Castree, N. (2004). 'Differential geographies: place, indigenous rights and "local" resources'. *Political Geography*, 23(1): 133–167.
Chakrabarty, D. (2000). 'Subaltern studies and postcolonial historiography'. *Nepantla: Views from South*, 1(1): 9–32.

Chandler, D. (2014). *Resilience: the governance of complexity*. London: Routledge.
Chandler, D. and Reid, J. (2016). *The neoliberal subject: resilience, adaptation and vulnerability*. London: Rowman and Littlefield.
Cutter, S.L. (2016). 'Resilience to what? Resilience for whom?' *Geographical Journal*, 182 (2): 110–113.
Davoudi, S., Brooks, E. and Mehmood, A. (2013). 'Evolutionary resilience and strategies for climate adaptation'. *Planning Practice and Research*, 28(3): 307–322.
Delanty, G. (2003). *Community*. London: Routledge.
Devadas, V. and Mummery, J. (2007). 'Community without community'. *Borderlands*, 6 (1): 21–32.
Ernwein, M. (2014). 'Framing urban gardening and agriculture: on space, scale and the public'. *Geoforum*, 56(1): 77–86.
Esposito, R. (2010). *Communitas: the origin and destiny of community*. Stanford, CA: Stanford University Press.
Givoni, M. (2016). 'Between micro mappers and missing maps: digital humanitarianism and the politics of material participation in disaster response'. *Environment and Planning D: Society and Space*, 34(6): 1025–1043.
Grayson, K. (2017). 'Capturing the multiplicities of resilience through popular geopolitics: aesthetics and homo resilio in Breaking Bad'. *Political Geography*, 57: 24–33.
Grove, K. (2013). 'Hidden transcripts of resilience: power and politics in Jamaican disaster management'. *Resilience*, 1(3): 193–209.
Grove, K. and Adey, P. (2015). 'Security and the politics of resilience: an aesthetic response'. *Politics*, 35(1): 78–84.
Grove, K. and Chandler, D. (2017). 'Introduction: resilience and the Anthropocene: the stakes of "renaturalising" politics'. *Resilience*, 5(2): 79–91.
Guattari, F. (1995). *Chaosmosis: an ethico-aesthetic paradigm*. Sydney: Power Publications.
Haraway, D. (2006). 'A cyborg manifesto: science, technology, and socialist-feminism in the late 20th century', in D. Harvey (ed.), *The international handbook of virtual learning environments*. New York City: Springer, pp. 117–158.
Heggem, R. (2014). 'Exclusion and inclusion of women in Norwegian agriculture: exploring different outcomes of the 'tractor gene'''. *Journal of Rural Studies*, 34(2): 263–271.
Hyndman, J. (2012). 'The geopolitics of migration and mobility'. *Geopolitics*, 17(2): 243–255.
Ichinkhorloo, B. and Yeh, E.T. (2016). 'Ephemeral 'communities': spatiality and politics in rangeland interventions in Mongolia'. *Journal of Peasant Studies*: 1–25.
Jha, M. (2010). 'Community organization in split societies'. *Community Development Journal* 44(3): 305–319.
Kepe, T. (1999). 'The problem of defining "community": challenges for the land reform programme in rural South Africa'. *Development Southern Africa*, 16(3): 415–433.
Langman, L. (2005). 'From virtual public spheres to global justice: a critical theory of internetworked social movements'. *Sociological Theory*, 23(1): 42–74.
MacQueen, K.M., McLellan, E., Metzger, D.S. et al. (2001). 'What is community? An evidence-based definition for participatory public health'. *American Journal of Public Health*, 91(12): 1929–1938.
Massey, D. (2005). *For Space*. London: SAGE.
Mulligan, M. and Nadarajah, Y. (2012). *Rebuilding communities in the wake of disaster: social recovery in Sri Lanka and India*. New Delhi: Routledge.
Mulligan, M., Steele, W., Rickards, L. et al. (2016). 'Keywords in planning: what do we mean by "community resilience"?' *International Planning Studies*: 1–14.

Nancy, J.-L. (2000). *Being singular plural*. Palo Alto, CA: Stanford University Press.
Ong, A. (2005). 'Ecologies of expertise: assembling flows, managing citizenship', in A. Ong and S. Collier (eds), *Global assemblages: technology, politics, and ethics as anthropological problems*. Malden, MA: Wiley-Blackwell, pp. 337–351.
Parikka, J. (2015). *A geology of media*. Minneapolis, MN: University of Minnesota Press.
Pearson, L.J., Wilson, S., Kashima, Y. et al. (2016). 'Imagined past, present and futures in Murray–Darling Basin communities'. *Policy Studies*, 37(3): 197–215.
Pudup, M.B. (2008). 'It takes a garden: cultivating citizen-subjects in organized garden projects'. *Geoforum*, 39(3): 1228–1240.
Rancière, J. (2004). *The politics of aesthetics*. London: Bloomsbury.
Reid, R. and Botterill, L.C. (2013). 'The multiple meanings of "resilience": an overview of the literature'. *Australian Journal of Public Administration*, 72(1): 31–40.
Rose, N. (1996). 'The death of the social'. *Economy and Society*, 25(2): 327–356.
Sadri, A.M., Ukkusuri, S.V., Lee, S., Clawson, R., Aldrich, D., Nelson, M., Seipel, J. and Kelly, D. (2018). 'The role of social capital, personal networks, and emergency responders in post-disaster recovery and resilience: a study of rural communities in Indiana'. *Natural Hazards*, 90(3): 1377–1406.
Sandhu, K. (2016). 'A black feminist's dilemma'. *Feminist Dissent*, 1: 88–100.
Simon, S. and Randalls, S. (2016). 'Resilience and the politics of multiplicity'. *Dialogues in Human Geography*, 6(1): 45–49.
Staeheli, L.A. (2008). 'More on the "problems" of community'. *Political Geography*, 27(1): 35–39.
Watts, M.J. (2015). 'The origins of political ecology and the rebirth of adaptation as a form of thought', in T. Perreault, G. Bridge and J. McCarthy (eds), *The Routledge Handbook of Political Ecology*. London: Routledge, pp. 19–49.
Williams, R. (1976). *Keywords*. Oxford: Oxford University Press.
Zebrowski, C. (2015). *The value of resilience: securing life in the twenty-first century*. London: Routledge.

7
ADAPTATION MACHINES, OR THE BIOPOLITICS OF ADAPTATION

Kevin Grove and Jonathan Pugh

Introduction

Research on resilience has reached a paradox of sorts. On one hand, advocates for participatory adaptation and resilience programming assert that these interventions can empower marginalised peoples. Proponents argue that even though much conventional work on adaptation and resilience directs attention away from political, economic and structural determinants of vulnerability, they are no less political, for they now focus on vulnerable people's individual and collective agency. Participatory interventions, we are told, remove the psychological and cultural barriers to adaptation, and thus enable individuals to become active agents in their own vulnerability reduction and capacity building (Brown and Westaway, 2011; Aitken et al., 2011; Pelling, 2010). On the other hand, a growing wave of critical research has shown that, these good intentions notwithstanding, the *effect* of adaptation and resilience programming is often to depoliticise vulnerability and reduce adaptation to merely surviving the after-effects of neoliberal development and the environmental transformations it entails. Resilience initiatives construct subjects who live with vulnerability and insecurity, rather than challenge the uneven socio-ecological relations that produce this vulnerability (Duffield, 2011; Reid, 2012).

At the heart of this paradox lies a problematic modernist political imaginary that limits how researchers can envision what politics is, where it is located and who (or what) takes part in politics as such (Chandler, 2014; Pugh, 2014). A modernist political imaginary is founded on a sovereign subject of rights, interests, responsibilities and agency, which are ultimately reducible to the individual's will; politics is a struggle between competing wills (Pugh, 2009a). In this view, politics is something to be managed and negated through designing more effective governance mechanisms that will produce the most just and equitable outcomes possible. This imaginary makes the political project of much adaptation and resilience research paradoxically reliant on the institutions it seeks to transform: the subject to be empowered through more inclusive adaptation and resilience programming is constituted as such only through its relation with institutions such as the state, sovereignty, property and the social contract (Grove, 2013a). The subject to be empowered is thus constitutively incapable of doing anything beyond *reforming* these institutions, and thus securing their persistence in the face of social and ecological vulnerabilities they create in the first place (Evans and Reid, 2013).

Building upon previous work by the authors (Grove and Pugh, 2015; Grove, 2012, 2014a, 2014b), this chapter attempts to move beyond this paradoxical politics through elaborating the concept of *adaptation machines*. This concept enables us to think of environmental politics not through the figure of a sovereign subject, but rather through unsteady assemblages comprised of *intensive relations* between material and enunciative elements, and *diagrammatic interventions* that attempt to structure these relations in particular ways (Deleuze and Guattari, 1987). Intensive relations refer here to affective atmospheres that envelop extensive bodies – individual and collective, human and non-human – and give them capacities to respond to stimuli and act in the moment of becoming. Diagrams are ideal systems of relations between these extensive elements; they attempt to shape the affective atmospheres and thus control the kinds of actions that are possible. Thus, and as is detailed below through examples of fieldwork in Jamaican disaster management, the concept of adaptation machines directs attention to an ontologically prior politics of affect whose stakes are the production of subjectivity. This helps us see the modern subject of rights, responsibilities, interests and disempowered agency as a product of a biopolitics of adaptation that attempts to shape the possibilities for individual and collective life in emergent and interconnected socio-ecological milieus.

To demonstrate the utility of 'adaptation machines' for resilience, the next section reviews recent research on participatory adaptation and resilience, and defines what we mean by 'adaptation machines' in detail. This concept is then used to analyse the biopolitics of catastrophe insurance in Jamaica. An extended conclusion, which also includes a discussion of Félix Guattari's (1995) useful work on ethico-aesthetic paradigms of engagement, considers how this concept might contribute to a revitalised and radicalised politics of resilience.

Re-politicising Adaptive Capacity

Like all fields of knowledge, the truths produced through research on disaster management and climate change adaptation reflect underlying force relations (Davoudi et al., 2013; Grove, 2014c; see also Foucault 1994, 2003). The current interest in participation and reflexive governance (to be discussed further later) has been shaped by an ongoing confrontation between radical approaches in disaster studies and more conservative techno-managerial impulses. In the 1970s, groundbreaking essays in disaster management combined Freierian pedagogy and Marxist political economy to advocate for participatory disaster management programming that fundamentally altered state–capital–society relations (e.g. O'Keefe et al., 1976; Wisner et al., 1977; Maskrey, 1989). However, during the 1990s, this work was largely depoliticised as researchers began operationalising vulnerability (see Middleton and O'Keefe, 1998). More recently, the incorporation of resilience thinking in both disaster management and climate change adaptation studies further solidified this depoliticisation (see Chandler, 2012; Reid, 2012; Grove, 2013b, Grove, 2014c; Pugh, 2014). Studies on resilience and adaptation often reduce politics to an isolated sphere of action to be regulated through adaptive management techniques. Indeed, where politics receives explicit attention, researchers tend to focus on either *procedural* struggles, which sees formal politics as an institutional constraint on adaptive capacity, or *distributional* struggles over the costs and benefits of vulnerability reduction and risk mitigation. In both cases, politics is something that can be controlled through designing and reflexively monitoring more effective governance arrangements. This managed negation of politics allows empowerment to occur. Empowerment results from techno-managerial processes of adaptive management, participatory programming and reflexive governance, which gives marginalised people the power to develop and implement their own adaptation and resilience-building initiatives.

However, as indicated above, this reading of empowerment relies on a narrow understanding of politics limited by an implicit modernist political imaginary that assumes individuals are proprietors of rights, interests and agency that can be more or less constrained by other individuals. This imaginary is evident in two key claims in the literature. First, calls for governance reform designed to expand participation implicitly assume that actors are rational subjects who can unproblematically communicate their interests to one another. Because participation allows the self to recognise itself in the other (see Pelling, 2010, page 63), verbal exchange flattens subjective differences into interests that can be transparently communicated and managed through collective negotiation. This somewhat reflects an older Habermasian tradition of intersubjective rationality in participatory approaches to planning more generally (Forester, 1988; Healey, 1997). Second, the frequent assertion that participation gives actors more power to define their adaptation choices is founded on an understanding of power as the negative oppression of one will by another will. Empowerment here involves removing the constraints, both psychic and institutional, that limit the ability of individuals to define their adaptation choices and realise their outcomes (Brown and Westaway, 2011).

Critical scholars working from diverse theoretical angles have increasingly argued that the depoliticisation of environmental governance often results in adaptation and resilience initiatives that have the effect of reinforcing, rather than challenging, institutions of neoliberal rule that create vulnerabilities in the first place (Chandler, 2014; Pugh, 2014). Not only does resilience direct attention away from the political-economic roots of vulnerability (Gaillard,

2010; Cannon and Müller-Mann, 2010), it also creates risk-managing subjects who live with rather than challenge environmental insecurity (Reid, 2012). Here, these critiques are built on by advancing an alternative political imaginary for analysing disaster management and resilience, one based not on the figure of the sovereign subject, but rather on 'adaptation machines' (Grove, 2014a). This extends our more general thinking on participation as an assemblage (Grove and Pugh, 2015). Specifically, for us adaptation machines can be thought as a socio-ecological form of Deleuze and Guattari's (1987) 'war machines'. They are decentred, rhizomatic assemblages that immanently act to shape the possible adaptations populations may pursue in emergent socio-ecological milieus. As assemblages, they are comprised of affective relations that striate, bodies, desires, statements, discourses and so forth, and diagrams of power, or strategic interventions that attempt to shape how these bodies can relate to one another. Diagrammatic interventions are not reducible to the interests of sovereign agents; instead, they are self-organised responses to *problems*, or the affective tensions that result from bodies juxtaposed with one another (Deleuze, 1990). Indeed, there is no sovereign subject within adaptation machines, for the interests and agency often imputed to the individual's sovereign will are instead recognised here as products of the pre-individual affective relations that comprise assemblages (Bennett, 2007).

Because it foregrounds pre-individual, affective relations, the concept of adaptation machines enables us to understand environmental governance *not* as a depoliticised and technical process of managing competing interests, but rather as a machinic assemblage that attempts to intervene in and parasitically control life's vital forces. Guatarri's (1995) elaboration of machines as diagrammatic and autopoietic can help clarify this point. As diagrammatic, machines exist independently of the specific elements that make them up. Instead, they consist of ideal, deterritorialised relations between parts. Machines are thus always supplemented by external bodies, institutions, desires and enunciations that are drawn together into machinic assemblages. As autopoietic, machines are self-organising: they do not rely on a master subject for their formation or maintenance. Instead, they produce subjectivities. This is an aesthetic production: machines produce spacings and timings that segment a chaotic, immanent field of intensive relations, and orient bodies as subjects within this field. Adaptation machines can thus be thought as a specific kind of machine. They constitute a series of relations that segments a chaotic, immanent and immeasurable present into extensive entities such as self-organising socio-ecological systems, uncertain futures that carry the threat of catastrophic systemic collapse and risk-bearing subjects – individuals, communities and states. They produce these effects by operating on affective relations that comprise everyday existence. Adaptation machines turn the vital force of life – the 'adaptability' of the present, or the ability of people and things to become different in response to affective pressures – into the source of, and solution to, environmental insecurity in contemporary capitalist order (Grove, 2014a).

Thinking environmental governance through adaptation machines does not therefore dissolve politics into the techno-managerial pursuit of socio-ecological transformations, but rather recognises how adaptation and resilience initiatives are inherently biopolitical. This means that they attempt to fold life into calculated programmes of governmental control. The 'life' they target is understood as a complex, interconnected and self-organising system sustained through the circulation of information between social and ecological entities (Grove, 2013c). Adaptation and resilience attempt to govern these systems through diagrammatic interventions that manipulate the *quality of relations* between these elements – the life-world or the affective milieu in which humans and non-humans experience and adapt to emergence (cf. Massumi, 2009; Anderson, 2012). However, these attempts are not always successful: because adaptation machines are reliant on an external supplement, as above, they possess no adaptive capacity of their own. Instead, they operate through a variety of techniques that parasitically capture, constrain, direct and, in the extreme, negate adaptive capacity that inheres in affective relations between bodies and their surroundings (cf. Hardt and Negri, 2000). And yet, because this is an external relation, the *possibility* still persists for people and environments to respond to change and surprise in unexpected and uncontrollable ways. Understood in this way, a biopolitics of adaptation points to both the powers *over* life exercised through climate change and disaster governance, and the powers *of* life that constantly exceed and confound these interventions (Anderson, 2010; Grove, 2014a).

The concept of adaptation machines thus helps us recognise how adaptation and resilience initiatives are sites at which a variety of techniques, knowledges, rationalities and forms of power may be (re)configured around the ethical problem of how to live within a spatially interconnected and temporally emergent world. These are sites of ethical and political possibility, where life's constitutive powers – that is, the power to create new worlds, new subject positions and new systems of value that do not fit into the existing order of things – can be both affirmed and negated. To illustrate this biopolitics, the following section draws on a case study of a specific adaptation machine at work in Jamaican disaster management.

Catastrophe Insurance

Over the past decade, resilience researchers have increasingly recognised the potential for insurance as a climate change adaptation mechanism (Lorenzoni et al., 2005; Kunreuther and Linnerooth-Bayer, 2003). Insurance ostensibly enables states and individuals to access new sources of adaptation financing in the wake of climate change-related incidents such as increasingly common hurricanes or flooding (Bals et al., 2006). The 2007 launch of the Caribbean Catastrophic Risk Insurance Facility (CCRIF) marked a major step in developing an insurance mechanism related to climate change. CCRIF is a regional insurance facility that pools the catastrophe risks of 16 member states and transfers these risks to global reinsurance and financial markets. Pooling decreases the cost of coverage for governments by 40 percent, which allows them to insure their critical infrastructure against hurricanes and earthquakes (World Bank, 2007). This infrastructure, such as government buildings, hospitals, transportation networks and utilities, sustains the circulation of people, goods, finance and information that enables economic growth and development.

Foucauldian scholars have long recognised that insurance is a technique of liberal rule that governs through uncertain futures (Ewald, 1991; O'Malley, 2004). It deploys actuarial techniques to turn an uncertain future into calculable, commodifiable and transferrable risks, and create subjects who experience their everyday lives in terms of danger and risk (Foucault, 2008; Dillon and Lobo-Guerrero, 2008). The risk-filled worlds insurance creates, and the risk-managing subjects that inhabit these worlds, are machinic effects of what Mitchell Dean (2004) calls 'risk assemblages'; networks of insurers, reinsurers, risk-management experts and state and international agencies that mobilise risk technologies such as catastrophe insurance.

Key for our concerns here is how CCRIF links the biopolitics of insurance with new state security rationalities. Member states' concerns with securing critical infrastructure against surprise disruptions reflects what Collier and Lakoff (2008) call rationalities of 'vital systems security'. This is a distinct vision of resilience and security that problematises the quality and integrity of infrastructure systems that sustain the circulations that support modern life. Indeed, CCRIF emerged in response to state fears of disorder that could result from its inability to repair critical infrastructure after a disaster. The driving force here was the so-called liquidity gap that threatened the Grenadian state's fiscal solvency in the wake of 2004's Hurricane Ivan. Ivan caused economic losses of US$2.4 billion in Grenada, more than double the island's GDP. The massive losses coupled with the storm's destruction of state infrastructure financially crippled the government and forced it to delay relief and recovery efforts for weeks (CCRIF, 2008). The fact that disorder did not occur in Grenada despite the state's fiscal insolvency did not deter regional leaders from imagining worst-case scenarios and, on this basis, lobbying the World Bank to develop a catastrophe insurance facility (Grove, 2012).

To provide Caribbean states quick access to capital in the wake of a disaster, CCRIF uses parametric insurance coverage. Parametric insurance is distinct from traditional indemnity insurance in that it does not cover all losses, but rather indexes a predetermined remuneration amount to the occurrence of specific weather events (Lobo-Guerrero, 2010).[1] In the case of CCRIF's hurricane coverage, the relevant parameters are wind speed and distance from a particular measuring point. The price of parametric insurance is determined not by actuarial methods that calculate risk on the basis of past disaster occurrences, but rather what Collier (2008) calls 'enactment-based' forms of knowledge that calculate and price catastrophe risk through models and simulations. CCRIF's coverage uses a five-layer catastrophe model to project the likelihood and resulting economic impact of disaster events – in this case, hurricanes with specific parameters (Grove, 2012).[2] In brief, CCRIF's models index state losses to wind speed and distance, and use computer simulations to predict the probability of storms with specific parameters and price the resulting catastrophe risk accordingly.

The ability of member states to access capital from global financial markets thus depends on CCRIF's catastrophe modelling. This turns the localised, territorialised insecurities that disasters pose to states into objective and deterritorialised catastrophe risks, that can then be transferred from states to markets through parametric insurance contracts, reinsurance and weather derivatives. The production and circulation of catastrophe risks blends together disparate actors into an adaptation machine of catastrophe insurance, a series of relations between, inter alia, critical infrastructure systems, communities, their biophysical surroundings, member states, catastrophe risk, CCRIF, risk management knowledge, catastrophe models, desires for security, reinsurers and global markets. Each entity derives its meaning from its relation to other entities within this series. Catastrophe modelling arranges these various material and enunciative components in a way that attempts to enable member states to respond to disasters more effectively and prevent disorder that threatens state security and capital accumulation. For example, individuals and communities become threats to state security; states become the subject of adaptation and environmental security; critical infrastructure is the object to be secured through catastrophe insurance; financial markets become the source of security; and the production and exchange of catastrophe risk through catastrophe modelling becomes the medium through which security can be achieved.

However, this serial arrangement depends on nothing more or less than the ability of catastrophe modelling to parasitically appropriate the constitutive adaptive capacity of individuals and communities marked as threats. While the *price* of catastrophe risk may be set through modelling and insurance contracts, the *value* of catastrophe risk derives from the potential difference it represents (Cooper, 2010). Specifically, catastrophe risk objectifies state fears of *possible* disorder that *could* result if states are unable to quickly repair critical infrastructure (Grove, 2012). It does this by projecting probable economic losses on the impact of a disaster, which negatively represents the state's ability to repair its infrastructure and regulate its population: the higher the loss, the more likely disorder becomes. But the source of disorder itself lies in the population's constitutive power to create difference, its ability to adapt in unexpected ways that derives from pre-individual affective relations. This intensive force saturates the present as the virtual potential for difference that may or may not be actualised as a disaster unfolds. For instance, communities affected by extreme weather events may not wait for relief supplies and instead may begin looting (undermining the institution of property); they may spontaneously organise disaster-response activities based on principles of self-help and mutual aid rather than following emergency response procedures (undermining state authority); in the extreme they may organise against the state if it fails to respond appropriately (undermining social order).

We can begin to see here the machinic and biopolitical effects of CCRIF's catastrophe insurance. Catastrophe insurance creates a world of (state) insecurity and vulnerability that can be managed by purchasing catastrophe insurance against the threats disasters pose to critical infrastructure. It creates risk-bearing subjects and member states and 'empowers' them to become more resilient to climate change – specifically, more frequent and intense hurricanes – through the purchase of catastrophe insurance. But this adaptation machine runs on the adaptive capacities of people and environments vulnerable to environmental change and surprise. Through techniques of catastrophe modelling, it deterritorialises the adaptive capacities of people impacted by a disaster from its localised affective milieu, and reterritorialises them as catastrophe risk within circuits of global finance. As such, the adaptation machine of catastrophe insurance attempts to consolidate neoliberal order by turning the threat adaptation poses to state order into a value member states can leverage on global markets in order to more effectively regulate their populations during disaster response. It reconfigures disaster management, not around the goal of reducing human suffering, but rather to pre-empt the threat that human suffering poses to state-based order.

As an adaptation machine, catastrophe insurance in the Caribbean assembles a loose and contingently aligned network of institutions, communities, individuals, measurement devices, computer programs and the biophysical environment through techniques of catastrophe modelling, risk pooling, reinsurance and derivatives trading. As in Deleuze and Guattari (1987), this process of assembling is driven not by subjective interests, but by affective relations – the affects that circulate amongst people and their surroundings before, during and following a disaster event, and state fears of the capacities for change that these affects create. CCRIF's catastrophe insurance appropriates this vital force through modelling techniques that negatively code the population's world-forming adaptive capacities as catastrophe risk that can be traded on financial markets for the security of both capital accumulation and the state order in an uncertain socio-ecological milieu.

Conclusions

This chapter has used the concept of adaptation machines to critically interrogate the depoliticisation of environmental governance. Through an example of adaptation machines in catastrophe insurance in Jamaica, it shows that the current focus on governance reform as an antidote to procedural and distributive injustices that disempower local peoples covers over an ontologically prior biopolitics of adaptation. This biopolitics works on an affective level to position people, communities and their biophysical surroundings within machinic assemblages, and thus produce the objects and subjects of neoliberal environmental security: critical infrastructure systems susceptible to disruption during a disaster event, hyper-adaptive communities that can threaten state order, communities that spontaneously 'properly' adapt to disasters and environmental change without external intervention and so forth. As such, this concept helps us locate the depoliticisation of resilience in a politics of affect that attempts to immunise neoliberal order against the external adaptive capacities that threaten the possibility, however unlikely, of radical change or disorder.

While this line of critique may make 'adaptation machines' appear to be all-encompassing and totalising apparatuses that disseminate control throughout the affective relations that comprise everyday life, the concept of adaptation machines also points to possibilities for more radical kinds of adaptation and resilience. Key here is how this concept enables us to see adaptation as neither an end goal nor a process that strives towards greater sustainability and equity (see Schipper and Pelling 2006), but rather what comprises life: to adapt is to live, to affect and be affected by others (social and ecological, material and immaterial) in the immanent moment of becoming. As such, the concept of adaptation machines highlights a largely overlooked ethical and political terrain. Because adaptive capacity inheres in affective relations, it is not something individuals, communities and states can lack. It cannot be increased or freed through techno-managerial interventions designed to increase social capital, engineer cultures of safety or empower marginalised populations (cf. Brown and Westaway, 2011). Instead, these interventions can only ever appropriate target populations' adaptive capacities and direct them towards certain adaptive actions and socio-ecological outcomes rather than others. The externality of adaptive capacity always leaves an opening around the question of *how* adaptive capacity can and should be appropriated in the name of responding to climate change and building resilience to social and ecological insecurities.

This question cannot be reduced to competing interests of sovereign subjects that can be rationally communicated and managed through improved governance mechanisms, because the sovereign subject with interests, rights and agency is a machinic effect of governmental interventions that attempt to create life-worlds and the subjects that inhabit these life-worlds. Instead, at the heart of this question lies what Peter Adey (2012) calls an 'ethos of assemblage': recognising the self as constituted out of, and as part of, overlapping and potentially contradictory assemblages, and as a result, an openness to being affected by and affecting others, in unplanned, surprising ways. This ethos can point towards radical collaborative research that is not organised around the problem of securing an ideal and deterritorialised 'life' against climate change and environmental surprise. The problem here, following Esposito (2013), is that the drive to secure an abstract human (or systemic) life is itself an aesthetic practice that detaches this life from the spatialities and temporalities of the lives it seeks to protect. This detachment is evident in adaptation and resilience programming that problematises, appropriates, controls and negates immeasurable adaptive capacity. In contrast, an ethos of assemblage immerses life in the intensive relations that comprise actually existing vulnerability, suffering and insecurity. It values adaptive capacity not as a threat to systemic persistence, but rather as the vital force of social and ecological change that *can* be appropriated and directed towards subversive and radical forms of resilience, rather than reactionary ones. The collaborations it advances are not structured by the concepts of disaster management, resilience theory, hazard studies or any other institutionalised system of knowledge that reduces adaptive capacity to a measurable and quantifiable entity. Instead, they are groundless collaborations that radically deserialise resilience and adaptation – that disconnects them from the rationalities and techniques of neoliberal environmental security and reconfigures them around a shared ethic of reducing insecurity, vulnerability and suffering, *however collaborators may define and sense these categories*. As part of adaptation machines that cannot help but appropriate resilience, researchers on climate change adaptation and disaster management have the potential to incite, and take part in, radical performances of resilience and adaptation that destabilise rather than secure the political economic inequalities that create contemporary environmental insecurities.

Here then, thinking about future directions, a good source of inspiration that has been found increasingly useful in our research on resilience is Félix Guattari's (1995) *Chaosmosis*. This book was written both as a response to the increasing chaos and disruptions of late capitalism, and to the hardening machinic effects of governmental apparatus (such as those resilience machines discussed above) as they seek to intervene in life, society, economy and environment. In contrast to traditional paradigms in the sciences, social sciences and humanities, in *Chaosmosis*, Guattari (1995, page 10) instead encourages the experimental development of what he calls new 'ethico-aesthetic paradigms'. Concerned with how subjectivity is produced, captured, enriched and reinvented through assemblages, for our specific interest in resilience machines in this chapter ethico-aesthetic paradigms direct our analytical attention in two key directions: (1) toward the contextually specific affective relations that permeate assemblages and create desires and capacities, such as desires for better resilience, participation, development or risk management; and (2) to diagrams of power that attempt to direct these affective relations and channel the constitutive power of bodies coming together toward certain outcomes rather than others. Together, these concerns draw out the ethico-aesthetic potentialities: participation in resilience programmes or projects involves a variety of techniques that bring desires, institutions, people, things and knowledge together in ways that can consolidate existing ways of life, or create entirely new possibilities. Here, as explained previously, more traditional forms of modernist resilience and participation can be seen to enact a negative biopolitics that reduces constitutive power, the force of difference, to substances – voice, social capital, resilience capacity – that can be controlled and regulated in ways that do not necessarily threaten institutions such as sovereignty that produce inequality and vulnerability. By contrast, a more experimental ethico-aesthetic paradigm of resilience would instead foreground the transversalising relations that permeate participatory assemblages, the specific social, cultural and technical machines that mediate these relations, and the creative potential for more innovative refrains that do not harden but become generative of a more affirmative biopolitics of life. This way of conceptually approaching debate in terms of ethico-aesthetics has been found useful in our practical and engaged research on resilience.

For example, in 2003 an initiative paying 128 Caribbean fisherfolk to explore the constraints and opportunities for their own fishing communities was successfully pitched to the British High Commissions of seven Caribbean islands by a group of Caribbean fisherfolk and one of the authors of this paper. The central rationale was that fisherfolk across the Caribbean should be their own paid development consultants, rather than have others do the job on their behalf. As their own consultants, fisherfolk would explore the nature of different fishing communities, how they could better develop fishing networks, new forms of political and social subjectivity across the eastern Caribbean. This experimental approach has good grounding in the Caribbean context. As the Caribbean writer Édouard Glissant (1997) says in his seminal 'Poetics of Relation', the Caribbean is a vibrant region that developed co-extensively through the opening up of relations (see also Pugh, 2013b, 2016). Yet, at the same time, the reductive Hegelian master–slave relationship, associated with such contemporary forces as the top-down international development consultant as analyst, and the overcoding machines of the international development industry more generally, exhausts the possibilities for opening up new relations. By contrast, the purpose of the fisherfolk project was to foreground the creative potential for more innovative refrains that do not harden but generate more affirmative biopolitics of life. For the first few years of the project this did happen. Fisherfolk generated a range of new connections and solutions around dozens of concerns they identified themselves (from how to mitigate conflict over fishing resources, to assistance after hurricanes and other disasters). Whilst, at the same time, they were simultaneously showing that their project should be driven by fisherfolk and not international development consultants (see Pugh, 2013a for detailed analysis). However, as time passed, the potentiality of this wider participatory assemblage began to close down and harden, resulting in anxiety and fatigue for many involved. As the project developed, fisherfolk were often physically exhausted from travelling between different islands to spend the money before expenses were shut off by funders, whilst at the same time trying to maintain a family and fishing life of their own. Many fisherfolk found it difficult to psychologically get to grips with this bizarre funding logic that emphasised the importance of spending money quickly to meet some target or other, when surely it made more sense to be frugal and save money for another day, as much more work needed doing in future as well. Just as academics find the bureaucratic burdens of grant writing and reporting increasingly onerous, so did the fisherfolk who, as the driving force, were fully involved in this process. From the academic's perspective, too, serious moral choices then had to be made as the project came to an end: should he reduce the project in a way that generated income for the university rather than fisherfolk (despite its clear rationale and intentions)? Given the desire to challenge international development traditions a decision was made not to; but, as a result, the academic lost his job and had to look for work elsewhere. Although another university then employed him, combined with the concerns of the fisherfolk, the collective participatory assemblage hardened, causing exhaustion and fatigue. From the academic's side, university employers not only asked how much money the project was generating for the university, but also how we could data code its 'impact' when it was not being fully funded by the 'blue chip' funding stream of a research council. Would this project be an 'impact case study' in the Research Excellence Framework; and if not, is it 'worth' investing time in it, or, perhaps better, strategically reinvesting in writing papers to meet other league table metrics? Because of these and many other factors, despite the positive earlier years of this project, the possibilities were thus reduced for more ethico-aesthetic, experimental and transversalising approaches to emerge – reducing the potential for a more affirmative biopolitics of life.

Yet, we argue following Guattari that thinking about participation in terms of ethico-aesthetic possibilities does allow us to broadly characterise such innovative and experimental turns in participation more generally in terms of a distinct ethical comportment to the affective capacities participation can generate (for example, Cameron and Gibson, 2005; McCormack, 2008; Askins and Pain, 2011; Pugh, 2005, 2009b, 2013a, 2013b; Gerlach and Jellis, 2015). In cases such as the fisherfolk project, participation did not seek to emancipate marginalised peoples by giving them voice, or fulfil specific policy aims. Instead it sought to create newly affective spaces – physical, social and psychic – where sedimented powers might be circumvented or dislodged. In this case fisherfolk physically getting on planes and becoming development consultants themselves – leading the process and getting paid for it. There was thus more of a Guattarian emphasis upon 'renewed forms of sociality' (Bertelsen and Murphie, 2010, page 156). What we would then characterise as these more ethico-aesthetic paradigms are about creating new possibilities that might give rise to new kinds of affects and reflections. This takes us far beyond thinking about concerns such as resilience and community-based disaster management as merely technical and social machines that operate through techniques of subjection that corral affective potential in established and fixed identities. Instead, it expands the concept of participation in the resilience machine itself; so that it is situated adjacent to an affective incorporeality that opens out onto a more pathic subjectivity of creative processuality. It is a way of thinking about participation and resilience that does not build alternative futures from an understanding of power as something that is directed toward 'consensus' or 'being-against', but rather a creative process of reflexively becoming-otherwise (Osborne, 2009).

Notes

1 For example, the Jamaican government may purchase a US$70 million hurricane coverage policy. Its attachment point, the specific parameters that determine when the government receives a payment, may be set to a category 2 hurricane that is 50 km from one of the island's 13 measuring points, such as Norman Manley Airport in Kingston. The lowest parameter may trigger a 20 percent payout, or US$10 million. The payout then increases as wind speed increases and distance grows closer: a category 3 storm may generate a 40 percent payout; a category 4 event, 60 percent and category 5, 100 percent (Grove, 2012).

2 The first layer, the hazard module, uses historical data to represent the built environment's exposure to natural hazards. The second layer, the exposure module, analyses the value of the built environment exposed to a peril using a 900 m grid to classify the exposure of each block on the basis of the value of physical assets it encompasses. The third and fourth layers use data on building type and year built to calculate the cost to repair exposed assets after a catastrophic event, estimate the percentage of a structure's replacement costs that will be required to repair it after an event, and project the damage within each block that results from a storm of specific magnitudes. Finally, these models are run through thousands of simulations to generate probabilities for specific loss levels. For example, a loss of $1.5 billion may be attributed to a one-in-ten-year event, which has a one-in-ten chance of occurring within any given policy year (CCRIF, 2009).

References

Adey, P. (2012). 'How to engage? Assemblage as ethos/ethos as assemblage'. *Dialogues in Human Geography*, 2: 198–201.

Aitken, C., Chapman, R. and McClure, J. (2011). 'Climate change, powerlessness, and the commons dilemma: assessing New Zealanders' preparedness to act'. *Global Environmental Change*, 21: 752–760.

Anderson, B. (2010). 'Modulating the excess of affect: morale in a state of "total war"', in M. Gregg and G. Seigworth (eds), *The affect theory reader*. Durham, NC: Duke University Press, pp. 161–185.

Anderson, B. (2012). 'Affect and biopower: towards a politics of life'. *Transactions of the Institute of British Geographers*, 37: 28–43.

Askins, K. and Pain, R. (2011). 'Contact zones: participation, materiality, and the messiness of interaction'. *Environment and planning D: society and space*, 29(5): 803–821.

Bals, C., Warner, K. and Butzengeiger, S. (2006). 'Insuring the uninsurable: design options for a climate change funding mechanism'. *Climate Policy*, 6(6): 637–647.

Bennett, J. (2007). *Vibrant matter: a political ecology of things*. Durham, NC: Duke University Press.

Bertelsen, L. and Murphie, A. (2010). 'Félix Guattari on affect and the refrain', in M. Gregg and G.J. Seigworth (eds), *The affect theory reader*. Durham, NC: Duke University Press, pp. 138–161.

Brown, K. and Westaway, E. (2011). 'Agency, capacity, and resilience to environmental change: lessons from human development, well-being, and disasters'. *Annual Review of Environment and Resources*, 36: 321–342.

Cameron, J. and Gibson, K. (2005). 'Participatory action research in a poststructuralist vein'. *Geoforum*, 36(3): 315–331.

Cannon, T. and Müller-Mann, D. (2010). 'Vulnerability, resilience and development discourses in context of climate change'. *Natural Hazards*, 55: 621–635.

CCRIF (2008). *Annual report 2007–2008*. Grand Cayman: CCRIF.

CCRIF (2009). *Annual report 2008–2009*. Grand Cayman: CCRIF.

Chandler, D. (2012). 'Resilience and human security: the post-interventionist paradigm'. *Security Dialogue*, 43(3): 213–229.

Chandler, D. (2014). *Resilience: the art of governing complexity*. Abingdon: Routledge.

Collier, S. (2008). 'Enacting catastrophe: preparedness, insurance, budgetary rationalization'. *Economy and Society*, 37(2): 224–250.

Collier, S. and Lakoff, A. (2008). 'The vulnerability of vital systems: how "critical infrastructure" became a security problem', in M. Dunn and K. Kristensen (eds), *Securing the homeland: critical infrastructure, risk and (in)security*. London: Routledge, pp. 17–39.

Cooper, M. (2010). 'Turbulent worlds: financial markets and environmental crisis'. *Theory, culture and society*, 27(2–3): 167–190.
Davoudi, S., Brooks, E. and Mehmood, A. (2013). 'Evolutionary resilience and strategies for climate adaptation'. *Planning Practice and Research*, 28(3): 307–322.
Dean, M. (2004). *Governmentality: power and rule in modern society*. London: SAGE.
Deleuze, G. (1990). *The logic of sense*. London: Bloomsbury.
Deleuze, G. and Guattari, F. (1987). *A thousand plateaus: capitalism and schizophrenia*. Minneapolis, MN: University of Minnesota Press.
Dillon, M. and Lobo-Guerrero, L. (2008). 'Biopolitics of security in the 21st century: an introduction'. *Review of International Studies*, 34: 265–292.
Duffield, M. (2011). 'Total war as environmental terror: linking liberalism, resilience, and the bunker'. *South Atlantic Quarterly*, 110(3): 757–769.
Esposito, R. (2013). *Third person*. Cambridge: Polity Press.
Evans, B. and Reid, J. (2013). *Resilient life: the art of living dangerously*. Oxford: Polity.
Ewald, F. (1991). 'Insurance and risk', in G. Burchell, C. Gordon, and P. Miller (eds), *The Foucault effect: studies in governmentality*. Chicago, IL: University of Chicago Press.
Forester, J. (1988). *Planning in the face of power*. Oakland, CA: University of California Press.
Foucault, M. (1994). *The birth of the clinic*. New York: Vintage.
Foucault, M. (2003). *Society must be defended: lectures at the Collège de France, 1975–6*. New York: Picador.
Foucault, M. (2008). *Birth of biopolitics: lectures at the Collège de France, 1978–9*. New York: Picador.
Gaillard, J.C. (2010). 'Vulnerability, capacity and resilience: perspectives for climate and development policy'. *Journal of International Development*, 22: 218–232.
Gerlach, J. and Jellis, T. (2015). 'Guattari impractical philosophy'. *Dialogues in Human Geography*, 5(2): 131–148.
Glissant, E. (1997). *Poetics of relation*. Ann Arbor, MI: University of Michigan Press.
Grove, K. (2012). 'Preempting the next disaster: catastrophe insurance and the financialization of disaster management'. *Security Dialogue*, 43(2): 139–155.
Grove, K. (2013a). 'On resilience politics: from transformation to subversion'. *Resilience*, 1(2): 146–153.
Grove, K. (2013b). 'Hidden transcripts of resilience: power and politics in Jamaican disaster management'. *Resilience*, 1(3): 193–209.
Grove, K. (2013c). 'Biopolitics', in C. Death (ed.), *Critical environmental politics*. London: Routledge.
Grove, K. (2014a). 'Adaptation machines and the parasitic politics of life in Jamaican disaster resilience'. *Antipode*, 46(3): 611–628.
Grove, K. (2014b). 'Agency, affect, and the immunological politics of disaster resilience'. *Environment and Planning D: Society and Space*, 32: 240–256.
Grove, K. (2014c). 'Biopolitics and adaptation: governing social and ecological contingency through climate change and disaster studies'. *Geography Compass*, 8(3): 198–210.
Grove, K. and Pugh, J. (2015). 'Assemblage thinking and participatory development: potentiality, ethics, biopolitics'. *Geography Compass*, 9(1): 1–13.
Guatarri, F. (1995). *Chaosmosis: an ethico-aesthetic paradigm*. Sydney: Power Publications.
Hardt, M. and Negri, A. (2000). *Empire*. Cambridge, MA: Harvard University Press.
Healey, P. (1997). *Collaborative planning: shaping places in fragmented societies*. Vancouver: UBC Press.

Kunreuther, H. and Linnerooth-Bayer, J. (2003). 'The financial management of catastrophic flood risks in emerging economy countries'. *Risk Analysis*, 23(3): 627–639.

Lobo-Guerrero, L. (2010). *Insuring security: biopolitics, security, and risk*. London: Routledge.

Lorenzoni, I., Pidgeon, N. and O'Connor, R. (2005). 'Dangerous climate change: the role for risk research'. *Risk Analysis*, 25: 1387–1398.

Maskrey, A. (1989). *Disaster mitigation: a community based approach: Oxfam Development Guidelines No.3*. Oxford: Oxfam.

Massumi, B. (2009). 'National enterprise emergency: steps toward an ecology of powers'. *Theory, Culture and Society*, 26(6): 153–185.

McCormack, D. (2008). 'Thinking-spaces for research-creation'. *Inflexions* 1(1): 1–16.

Middleton, N. and O'Keefe, P. (1998). *Disaster and development: the politics of humanitarian aid*. London: Pluto Press.

O'Keefe, P., Westgate, K. and Wisner, B. (1976). 'Taking the naturalness out of natural disasters'. *Nature*, 260: 566–567.

O'Malley, P. (2004). *Risk, uncertainty, and government*. London: Glasshouse.

Osborne, A. (2009). 'The ethico-aesthetic paradigm and tribal assemblages and space rock'. Available at: https://totalassaultonculture.wordpress.com/2009/01/23/the-ethico-aesthetic-paradigm-tribal-assemblages-and-space-rock/, accessed 11 November 2016.

Pelling, M. (2010). *Adaptation to climate change: from resilience to transformation*. London: Routledge.

Pugh, J. (2005). 'The disciplinary effects of communicative planning in Soufriere, St. Lucia: governmentality, hegemony and space-time-politics'. *Transactions of the Institute of British Geographers*, 30(3): 307–321.

Pugh, J. (2009a). 'What are the consequences of the "spatial turn" for how we understand politics today? A proposed research agenda'. *Progress in Human Geography*, 33(5): 579–586.

Pugh, J. (ed.) (2009b). *What is radical politics today?* Basingstoke: Palgrave Macmillan.

Pugh, J. (2013a). 'Speaking without voice: participatory planning, acknowledgment, and latent subjectivity in Barbados'. *Annals of the Association of American Geographers*, 103(5): 1266–1281.

Pugh, J. (2013b). 'Island movements: thinking with the archipelago'. *Island Studies Journal*, 8(1): 9–24.

Pugh, J. (2014). 'Resilience, complexity and post-liberalism'. *Area*, 46(3): 313–319.

Pugh, J. (2016). 'The relational turn in island geographies: bringing together island, sea and ship relations and the case of the Landship'. *Social and Cultural Geography*, 17(8): 1040–1059.

Reid, J. (2012). 'The disastrous and politically debased subject of resilience'. *Development Dialogue*, 58: 67–79.

Schipper, L. and Pelling, M. (2006). 'Disaster risk, climate change and international development: scope for, and challenges to, integration'. *Disasters*, 30(1): 19–38.

Wisner, B., O'KeefeP. and Westgate, K. (1977). 'Global systems and local disasters: the untapped power of peoples' science'. *Disasters*, 1(1): 47–57.

World Bank (2007). *Results of preparation work on the design of a Caribbean catastrophe risk insurance facility*. Washington, DC: World Bank.

8

THE RESILIENT CITY: WHERE DO WE GO FROM HERE?

Peter Rogers

Resilience is a concept which, if used well, may help deliver a paradigm shift in how an urbanisation is governed in the 21st century, yet the term remains a focus of scepticism and critique for many who encounter it. This is in no small part due to the fragmentation of the concept, pulled apart and put back together again as vested interests seek to exploit the boom in markets opened by resilience to new disaster capital investments. The competition between emerging narratives also provides a foundation for multiple framings in strategy, policy and practice. Strategies often set the stage for organisations who then implement policy experiments attempting to deliver more resilient human settlements and, importantly, more resilient humans. Into this contested landscape surprisingly few investigations have attempted to unpack the complex interplay of strategic policy with the lived experiences of those implementing these new ways of resilient working in practice, tending to focus on the concept (strategic definition) or the subjects (experts or citizens) but not explore the links between them. This is particularly important for urban resilience and the 'resilient city' because the ways in which resilience as strategy is *acted out by* people is at the heart of any potential conversion of an 'urban growth machine' into an 'urban resilience machine'. This chapter takes a first step into this gap by exploring two global 'urban resilience' policy frameworks, forging clearer links between resilience as rhetoric in strategy or policy and the ways of acting enabled by them. This effort opens up a more nuanced appraisal of resilience ethics amongst practitioners emerging from the ongoing interplay of urban growth, strategic policy and the *techniques of governing* used to implement more resilient cities.

The Challenges: Scale, Scope and Focus

There is a lot of ground to cover in this debate. Resilience appears as a strategic buzzword in many policy experiments targeting a host of wicked problems. What follows will touch on a number of strategies, but mainly highlights the need for better articulation of the bridge between strategy, policy and practice. It may be rather easy to consider resilience as debunked by critics who see it as no more than a new form of neoliberal discourse (Tierney, 2015), but this critique is one-sided. When unpacking the rhetoric of resilience it is easy to dismiss the focus on the self as a debilitating influence on political action – in so far as political action is seen to be based on the transformation of structural inequalities in terms of traditional political economy. This is hard to deny, but strategic framings of resilience and the potential governance of resilient subjects is not only about the citizen–state relationship, it also affects the *acting out* of professional conduct in the implementation of resilience, embedded as it must be in actions of resilience practitioners. A nuanced appreciation of how strategy is translated through the lens of those implementing it helps us to better understand both the technique of governance and the practitioner enacting that technique.

Changing this emphasis allows researchers to track how resilience thus shifts where the transformative change may occur; away from structural inequality to interpersonal actions. In response the focus of critique must move away from traditional structures of capital as *praxis* onto the interplay of beliefs, norms and rules of conduct as they are dynamically created and acted out, onto *poiesis*,[1] where the potential for change emerging from the process of governing resiliently can be better understood. Transformative potential is thus not embedded in the battle over a political economy of structural relationships – which capital must always win – but rather in the manner by which a policy is implemented. What techniques are used for the disposition and delivery of resilience, how they are acted out and how they thus (re)shape the sociomaterial processes and relationships enmeshed in a specific context becomes the central focus for future research. I argue that this work can be begun here through (a) analysis of specific strategies concerned with resilience – and the tools they prioritise – and (b) the application of these techniques of governing in particular urban settlements. Though the former takes precedence in this chapter and the latter is subject to further research, some implications for researching resilient cities can be traced out from what follows.

This approach changes the scale, scope and focus of analysis considerably, drawing attention to *resilience as process* and as a *technique of governing*. Contextually embedded techniques of governing may *appear* to have the goal of building resilience, but can also have unintended and, as yet, unknowable impacts on both settlements and subjects. The unintended consequences of change emerge already enmeshed in specific resilience projects (as policy experiments) tied to the idiosyncrasies of the places they are enacted and the actors who enact them. The process of resilience building may thus be seen as an ongoing interplay of contextually embedded experiments, transformative by nature but nuanced by context. Each experiment is encountered uniquely in each case, creating not a systematic method for optimising resilience so much as an ethical framing for how that work can be conducted to create opportunities for transformative change reassembled in new ways over and over again. This also creates opportunities for researchers to unpack the implications of urban resilience and the relational interplay which shapes both the city as a human settlement and the resilient subject as an inhabitant of it.

The Subject and the Strategy

A reflexive and engaged citizen-subject, as desired by resilience, has a direct stake and role to play in the enactment of urban governance. If resilience thinking is to become systemically embedded as a technique of governance then that also comes with a requirement for the individual subject to have greater *capacities to act* embedded in governance more directly. This is not a debate over the manipulation of citizenship, but rather the balancing of the problems resilience-led thinking can address – framed *in* strategic policy – with the ethics and conduct of those seeking to address the problems – embedded in methods and tools as techniques of governing guided *by* strategic policy. The subject is not just the citizen, but rather the stakeholder, which levels the playing field between expert and lay person by focusing on their mutual capacity to act on an issue of concern to both, thus opportunities for redefinition of the relationships in which they are engaged begin to emerge.

In the governance of a resilient city this is enmeshed in existing discussions and practices of urban growth, sustainable development, environmental management, climate adaptation and mitigation, disaster management, security and social order management (often discussed in relation to terrorism but also civil disobedience) and a host of other fields which interplay through encounters with resilience. Resilience as strategy is important because it helps frame the encounter with the resilient city. It also helps ethically ground the resilience thinking in particular forms of acceptable action, particular tools for delivery. These themselves are beginning to influence in perhaps unintended ways the emergent governance of the resilient city and the relationships between resilient subjects in differing contexts.

Current research on the resilient city has yet to fully appreciate the opportunities for counter-neoliberalisation inherent in such a reworking of the resilient subject. It is difficult to grasp that the interplay of ethical and political work on the self as undertaken by subjects 'becoming more resilient' is embedded in a system of resilience-led governance that itself is changing. The resilient citizen-subject requires greater collaborative influence, must participate more, but equally so must experts. This has the potential to alter the balance of power in micro-relationships across a broader sense of 'civil society'.[2] The consequences of such change, and wider adoption of resilience-led ways of working, are not yet known. Neither do we know which techniques of resilient governance will emerge dominant. It is, however, likely that whichever techniques of resilient governing are 'mainstreamed' will be driven by the organisational coalitions most successful in inculcating *their* view of resilience into the work of practitioners, who do the hard graft in delivering outcomes.

To this end this chapter cannot offer a final analysis, rather it opens up the possibility for rethinking how we study resilience in human settlements. It explores the need for better understandings of the links between strategic framing, problem statement, project management, latent capabilities of stakeholders (including the public), the expertise of specialists and lay persons, organisational cultures and established methods of workings as well as market forces, which drive many conditions of urban growth (from resource exploitation to services delivery and access) – all must be considered when assessing the urban resilience machine. Who is to be given access to which levels of decision making, what links from strategy to practice emerge, which coalitions dominate is by no means clear. Some organisations emphasise expert knowledge only, others implement resilience via techniques of collaborative community building. Many conflicting approaches encourage different forms of resilience-oriented behaviours, amongst both professionals and lay persons. International non-governmental organisations are in negotiation with governments at different scales (national, regional and local). Private-sector interests, civil society organisations of all types, individual citizens and more, all have an interest and the stakes are high. It is only possible to map a part of these changes in this chapter, highlighting some tensions and possibilities. The international strategies selected are chosen because each frames resilience as a core business of urban governance, but remains unfinished with uncertain outcomes. Resilience as encountered in such strategies shows the scale and scope of the dilemma. Linking strategy to the process of implementing urban resilience identifies governance techniques, however at the time of writing many cities are working on their own resilience strategies and few have reached a full implementation so the outcomes remain uncertain. By focusing on the bridge of strategy, policy and practice as it is to be implemented in complex urban systems one can better appreciate the tensions between existing practices, dominated by economic growth, and those emerging from resilience-led strategy to better understand the challenges that resilience practitioners will face over the longer term.

Resilience in strategy often places less importance on the governance of *economic growth* than on the governance of *capacities of actors* across *all* components of a 'complex interdependent urban system'. Using this as a point of departure one can better gain an understanding of complexity in systematising the reflexive ethics of resilience as a *process* or *technique of governing*, perhaps at odds with the actuarial ways of working adopted by many organisations who rely on these 'top-tier' strategic frameworks to orient their actions and investments. One can both highlight the potential for a shift in governance, fuelled by resilience, as a challenge to the political economy of place so deeply grounded in the 'urban growth machine', as well as show the risks for this emerging way of working to become absorbed by the neoliberalising economic growth engine with which so many critics of capitalism are now familiar, but to do so one must go beyond theory and look at the rhetoric of resilience in strategies affecting human settlements before linking this to the ways of working that this privileges in the governance of the resilient city.

Urban Resilience in Global Policy

As the star of resilience has risen so have the range of strategies which seek to build a coherent framework for its operationalisation and implementation. The Sustainable Development Goals (SDGs) approach is where explicit links are made between climate change and urban resilience (United Nations, 2015a). With a 15-year lifespan this global policy platform, the SDGs, proposes 'to end poverty and hunger everywhere; to combat inequalities within and among countries; to build peaceful, just and inclusive societies; to protect human rights and promote gender equality and the empowerment of women and girls; and to ensure the lasting protection of the planet and its natural resources' (United Nations, 2015b, page 3). A key problem in enacting this ambitious agenda will be managing the sometimes 'toxic politics' of climate change, working across the emerging resilience agenda, sustainable development, climate mitigation and climate adaptation initiatives and the legacy of the Millennium Development Goals which the SDGs replaced (United Nations, 2015c). The most explicit link is made through Goal 11 for *inclusive, safe, resilient and sustainable cities and human settlements* which offers only the broadest use of resilience as a means to target projects for reducing inequality and improving quality of life amongst the most vulnerable inhabitants of cities subject to rapid growth.

Clearly resilience is not the central driver of the SDG agenda, but one component amongst many; for a more directed overview of nuance within resilience strategy there are no shortage of offerings. A range of programmes include the 'New Urban Agenda' from Habitat III, the United Nations (UN) 'Making Cities Resilient' campaign (United Nations, 2016) and 'third-sector' funding from the Rockefeller Foundation '100 Resilient Cities' programme (Rockefeller Foundation, 2016a), also facilitated by the socio-ecological traditions of the Resilience Alliance (Resilience Alliance, 2016) and Stockholm Resilience Centre (Stockholm Resilience Centre, 2016). Each organisation seeks to do good work within their organisational remit, but each also mobilises the idea of resilience in different ways. It is not the time nor place to revisit the exhaustive definitions debate to nail down specifically what resilience *means*, as each strategy offers subtle variations. Suffice for us to say that resilience is increasingly a *polysemic* concept (Reghezza-Zitt et al., 2012; Rogers, 2017) reappearing in each policy platform with a different genealogic encounter shaping the work and tools in play. This emphasis depends on the remit and goals of the relevant organisations as they mobilise their own agenda, funding schemes and projects, in order to meet different types of goals and targets.

Resilience projects do have common features, but similarities only add to the complexity as the different underlying logics require different understandings of the idea as well as different skills for those called to act upon and implement the work itself. Many strategies share a strong emphasis on self-determination for citizens – seeking to draw out existing capacities to act amongst the population in their local context (Attorney General, 2011; UK Resilience, 2005; Civil Contingencies Secretariat, 2004). Such national strategies have drawn criticism as the new face for expansion of biopolitical forms of neoliberal governance (Filion, 2013; Joseph, 2013). This is not surprising as there is a strong infrastructure and economic rationale for improved resilience affecting organisations across traditional sectoral boundaries, such as utilities, civil service, experts (from disaster managers to risk managers to emergency services). Even when well defined such national strategies can be subject to conflicts of interest between stakeholders when implemented in practice (Rogers, 2011). The coordination of resources is another significant issue within resilience-building debates (Christopher and Peck, 2004; Ponomarov and Holcomb, 2009). The vulnerability of 'just-in-time' system optimisation governance practices inherent to growth economies have increased the attraction of resilience-oriented approaches to technical operations in areas such as supply chain risk management (World Economic Forum, 2013), again focused on risk-averse practices driven by expert professionals governing at a distance.

Resilience is not, however, entirely dominated by such 'neoliberalising' discourses. Sustainable development contributes a more humanitarian orientation, towards improving the access of disadvantaged populations to basic needs – from water, sanitation, education and housing to sustainable economies for local products (United Nations, 2015b, 2015c); just as in the managerially oriented strategies this requires stakeholders to cut across traditional sectoral silos. Community development has also continued to raise the profile of the general public, leading to an increased profile for community-formation skills and practices amongst resilience workers seeking to build resilience by improving common understandings of needs and wants in local contexts (Mulligan et al., 2016). Traditional approaches to disaster management have also led to more nuanced metrics for the outcomes of different types of resilience projects (see for example Cutter et al., 2008; Lee et al., 2013; Magis, 2010). Such multi-agency and portfolio-crossing appeal requires cross-pollination of skills and broader thinking on the unintended benefits of resilience thinking in practice. The SDGs are an important global strategy platform used to help guide a number of other frameworks, but in relation to the emergent resilience discourse they are of limited use in understanding the techniques to be adopted by those implementing this agenda.

A single form of resilience cannot be locked down at the global or international level of analysis, demonstrating how the polysemic conceptual framing enables multiple encounters and resilience of different forms to emerge in different contexts. Because no one-shot resilience is in evidence there are no standardised tools or techniques for resilient governing of urban growth; rather, there is subtle standardisation implicit in cross-sectoral collaborations beginning to emerge from the strategic framing of resilience as a precursor to risk and needs assessment at a subnational level. Here the 'New Urban Agenda' and '100 Resilient Cities' offer more focus for resilience thinking to be grounded work done by resilience practitioners. By working at a subnational, regional and city-regional scale these policy platforms begin to bridge the strategy, policy, practice gap for those seeking to dig beneath strategic rhetoric and facilitate the translation of resilience from such rhetoric into techniques or tools that can be used to deliver resilience in practice and govern in ways that encourage a potentially measurable form of resilience over the longer term.

Habitat III and a 'New Urban Agenda'

Resilience does not feature at all in the first two Habitat accords which focused much more on the broader field of sustainable development in ways similar to the SDGs. However, it appears frequently in the 'zero draft' of the third accord. Again, like the SDGs, resilience is used broadly, emphasising the 'strengthened resilience' of cities and 'resilience building' as an ongoing activity. The zero draft places themes of inclusivity, safety and equity central to the third 'transformative commitment' of the 'New Urban Agenda', to 'Foster ecological and resilient cities and human settlements, driving sustainable patterns of consumption and production, protecting and valuing ecosystems and biodiversity, and adapting to and mitigating the impact of climate change while increasing urban systems resilience to physical, economic, and social shocks and stresses' (Habitat III, 2015, page 3).

The techniques of resilience building to be mobilised here appear most aligned with understandings of vulnerability, risk management and disaster risk reduction – used widely in development research (see for example Thomalla et al., 2006). Clearly there are also 'opportunities associated with the rising incidence and costs of urban disasters, the current and anticipated impacts of climate change, and the protection of critical ecosystem services and natural resources' (Habitat III, 2015, page 1). Such opportunities may manifest in disaster capital investments with variable outcomes (Pelling and Dill, 2010; Timms, 2011). There is a concern that the implementation plan developed under the aegis of Habitat III will afford little more than a rebranding opportunity for pre-existing programmes wishing to secure future funding support. This creates an implementation landscape where the drive for measurable and achievable actions focused on implementation and prior track record stymie the development of innovative solutions at the heart of resilience rhetoric; for when resilience is treated as an aspiration with certain operational requirements it remains little more than an amorphous framework similar to that of the SDGs above.

The 'New Urban Agenda' seeks a grounded framework to avoid this through the creation of a tactical model for resilience building which can be mobilised across local contexts by creating intersections between particular resilience types and particular hazard types. This is a common tactic for tracking the means by which different organisational coalitions seek to mobilise techniques of governing that align resilience rhetoric with the appropriate skills of key stakeholders used to dealing with local response to local events. A number of such typologies of resilience can be found in the disaster resilience literature itself closely connected to urban resilience research (Alexander, 2013; Coaffee et al., 2009; Godschalk, 2003). Analysis of governance undertaken via typology requires a nuanced appreciation of which fields are in play to target and mobilise skills appropriate to the challenge at hand. The 'New Urban Agenda' treatment uses this approach not for looking at the separate systems as components of urban resilience, but rather at the interplay between them across scales (see Figure 8.1), which is encouraging but still with its limitations being focused more on acute shock events (fire, flood, bomb, etc.) than slow-burning stressors (e.g. poverty).

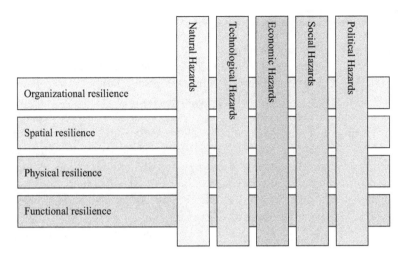

FIGURE 8.1 Urban systems model approach
Source: author's own, adapted from United Nations (2015a, page 3)

As noted previously, models and typologies have proliferated under the glut of interest in defining and operationalising resilience as an implantable strategy for policy (Handmer and Dovers, 1996; Klein et al., 2003; Rogers, 2015a) but consensus is not advisable, nor even possible. The strategic treatment of resilience as a framing concept for subthemes may yet be mobilised through more traditionally focused policy interventions, but the broader negotiations around Habitat III have opened up the conversation on governance at the local level much further than the SDGs. Of 22 issue papers within the Habitat programme paper, 15 explicitly discuss *urban resilience* in depth, particularly aligned with *urban ecology* and *environment*. These discussions also flag the vital issue of land-use management and tensions in public accountability to corrosive private-sector influences. The implementation tools levied argue for resilience 'action plans' to leverage resilient-led practices *within* existing planning arrangements and policies, but such tools may not alter the inherently corrupt land-use planning associated with development around the world. A more positive note, however, is present as an awareness of the systemic interdependencies of complex systems into governance is built into the discussion drafts, tied to 'developing mechanisms/instruments to promote coherence across systems, sectors and organizations related to their policies, plans, programs, processes and investments in urban resilience' (Habitat III, 2015, page 8). Despite the rhetoric supporting local agency ownership and coordination of planning decisions there is little within the agenda thus far to show how, where or when citizens themselves will become more active stakeholders.

The types of resilience presented by the 'New Urban Agenda' in its early stages here serve to lend a framing to different forms of micro-interventions that speak to distinct challenges posed by each hazard type, whilst generating an awareness of the interplay between these features of the complex urban system, but again remain necessarily broad and focused on expert stakeholders rather than manifesting evidence of the collaborative and participatory forms of resilience inherent in its potential. As the 'New Urban Agenda' moves from consultation to implementation it may be too early to criticise this. A grounding of implementable actions that turn resilience thinking into practice may yet still come forth. Some practical aspects of governance were on the table at the 2016 Quito meeting – for example, managing political pressures for decision making between government bodies across scales; drawing out more equitable, ethical and sustainable transnational corporation investments; increasing political and corporate accountability; or increasing access to participatory forms of budgeting and planning for more vulnerable and historically excluded populations – but resilience remains a hazard-driven, risk management-led form of governance in this iteration. Neither the SDGs nor the 'New Urban Agenda' appear to engage in sufficient depth with the technical challenges or opportunities to governing in a different way; the evidence for that comes from outside of the realm of these international policy platforms.

'100 Resilient Cities'

Both the UN 'Making Cities Resilient' campaign and the Rockefeller Foundation's '100 Resilient Cities' programme offer excellent examples of on-the-ground functional projects seeking to build resilience, however, despite the above strategic policy focus on UN initiatives, we will now turn to the Rockefeller foundations programme as an example of the influence of an international civil society organisation on the operationalisation and implementation of resilience in practice. The analysis here steps outside of the UN frameworks to demonstrate the difference between these initiatives and one led by a third-sector organisation to show how the implementation of resilience thinking interpreted elsewhere is generating a significantly different approach when linking strategy to practice.

At the core of the Rockefeller programme's approach to resilience is a different model of resilience, but one that still draws on a model of capacity building across work areas. This approach steps away from the usual expert actors and broadens the scope of stakeholders, significantly addressing 'the capacity of individuals, communities, institutions, businesses, and systems within a city to survive, adapt, and grow, no matter what kinds of chronic stresses and acute shocks they experience' (Rockefeller Foundation, 2016a). The model emphasises a bottom-up analysis of the challenges of assessing network capacities within a city governance structure particular to that locale.

The programme operates a direct intervention within governance tactics for delivering resilience by undertaking a competitive evaluation before a city is nominated for membership in the programme. Once evaluated by the

Rockefeller Foundation the scheme funds the appointment of a chief resilience officer (CRO) directly into a key governing organisation for that city – often the city council or equivalent body depending on the governance structure for the town or city. Who this appointee is, their area of expertise and where within the governance organisation they are located varied depending on the nature of the resilience challenges identified within the evaluation process. This may be a civil engineer specialising in flood infrastructure (e.g. in Bangkok) or a collaboration specialist from the non-governmental organisation (NGO) sector (e.g. in Sydney). Whilst the CRO is an officer of the governing body for that locale the CRO is also responsible to the Rockefeller Foundation, who supplies the funding for their operations. The goal of this appointment is to use tools and specialist consultancies recommended by Rockefeller to assess the resilience challenges of the locale, using pre-approved evaluation tools and consultants. Subsequent operations seek to then build an awareness of identified challenges into the core functions of government bodies in each member city, delivering a formal resilience strategy, including an implementation plan, within two years of the initial CRO appointment. The resilience strategy seeks to embed good planning practice for resilience building into the cross-sectoral governance for all areas of identified vulnerability, not limited to disaster (by itself) but rather emphasising networked capacities for leadership and management, economy and society, health and wellbeing and infrastructure and environment.

Where the SDGs follow a pattern of consultation established through the Millennium Development Goals and suffer from a poor public image and confidence regarding their 'implementability', and Habitat III seeks to build clear links between strategic guidance and practitioners but is still (at the time of writing) in the consultation phase and 'expert'-oriented, the '100 Resilient Cities' programme has taken a direct intervention in the creation of a network both in each city and outside of traditional civil society through their own network of CROs internationally. Networks thus circumvent traditional civil organisations' authority structures in some ways whilst working within them to embed resilience as core business, seeking to professionalise resilience practice without formal legislative support from nation states requiring this in quality standards or statutory conformance (Rogers, 2015b). This approach to the governance of resilience provides expertise grounded in a quality management culture familiar to risk-averse managerial forms of practice but targets identified challenges specific to that locale in ways that give multiple 'hits' on key performance indicators across departments that may be competing for column inches in the broader budgetary constrictions of austerity-led governance. A quality-led approach also creates a professional standard for practitioners ensuring that whilst operating within the politics of governance organisations the delivery of resilience is less likely to be swapped with a new policy buzzword. Multiple accountability structures that circumvent political rhetoric through professionalisation practices allows the CRO to remain partially insulated from organisational politics within which they and their teams are working. As such it creates a layering of responsibilities between traditional civil service operatives within formal government organisations, cutting across partnerships with the private sector across relevant fields (e.g. utilities, development, service provision and more), non-governmental agencies and community groups. It creates the relationships that facilitate greater collaboration and in so doing implicitly builds capacity within the broader structures of governance in ways that defy simple metric evaluation. This is a key feature of the technique of governance which creates the potential for more positive outcomes within the culture of organisations tasked with delivering resilience. The techniques of resilient governance require a shift in the underlying values away from risk aversion towards a collaborative empowerment of key skills where they appear within the network; something only possible if governance itself shifts its modus operandi away from risk management for system optimisation to risk coordination for flexible problem solving.

Research projects on the various resilient cities are now expanding, and the first resilience strategies are now beginning to emerge from some of the cities further along this path (Rockefeller Foundation, 2016b); though there is little holistic research on the impact of the programme in more global terms. In part this is still an ongoing programme – with the final inductees to the '100 Resilient Cities' coming only early in 2016 – but also as the range of foci across the emerging network of CROs is still very much an emergent system tied closely to the capacities, capabilities of the players in each city as well as the unique challenges in each local context. Rather than being driven from the top down by a global strategic vision, the '100 Resilient Cities' are inculcated within resilient practices from the bottom up using centrally devised ways of working that cut across established techniques and mechanisms of planning and policy implementation. The implications of this are not manifest in urban form, rather more significantly they manifest in the techniques of governing mobilised to build resilience thinking into the core functions of governance appropriate to each city where investment is made.

Implications of a Global Resilience Project

Each example one might choose from broader global strategies and then seek to link with specific local investments will draw out a slightly different element of resilience in practice. Each time we find different roles are being played out by organisationally diverse assemblages of civil society stakeholders, each building their own means of delivery appropriate to the urban resilience challenge most prominent in their own domain. This seems to suggest that it is the changing nature of work being done, and by whom, that shapes urban resilience. The changing methods and techniques of governing become more important than what resilience 'is' or 'means'. The way of working to deliver the projects is as important, if not more, than having a resilient transport network, building or infrastructure system. Civil society is increasingly 'comprised [of] a constellation of juxtaposed and changing elements that resist reduction to a common denominator, an essential core or first principle' (Jay, 1984, page 19). Resilience is not a 'first principle', rather an assemblage of both human and non-human elements, combined at strategic points of orientation. These points serve as points of departure for the mobilisation of skills and resources to enable targeted action, using the ways of working most appropriate and rolled out in smaller-scale policy experiments. It is a flexible form of governance, far more collaborative in its operation, but at odds with the rigid responsibility matrices of many organisations after 50 years of neoliberalism.

The resilient city is a fine crucible for the analysis of this process by which the changing techniques of governance associated with a given strategic orientation point can be understood, but a poor point of departure for the seeking of a grand narrative of resilient urban life. The diversity of organisational components which are at play in resilience-building activities varies wildly, as does the emphasis and potentially the significance of potential change. There is little in global policy (like the SDGs) to suggest that resilience is a 'game changer' in strategic terms. The implementation focus of Habitat III shows greater potential for changing the focus of governance towards resilience-led thinking than the SDGs, but whether this translates into an opening of access to groups traditionally excluded from decision making (e.g. low-income communities in urban slums) is yet to be seen. Yet this access must be opened if urban resilience is to be more than a machine for disaster capital-led urban growth.

In the resilience cities network emerging from Rockefeller's CRO model we see resilience as a strategic priority is directly inculcated into core governance practices. The strategic role is not one of setting the framework, but rather building the relationships and collaborative working ethics that link resilience thinking into the way governance organisations work. This collaborative emphasis is far closer to the understanding of resilience as a technique of governing rather than an outcome, but again at this stage the benefits of the programme itself are yet to be fully understood. Research that provides genealogical understandings of which type of resilience is in play and how it is understood by the actors are useful in showing what kind of actions may be taken, but limited in helping understand what means of working are being acted out in practice. Further research on resilience projects can provide case studies of best practice to expand the growing base for comparison of the techniques used by different operators. This is needed to help unpack the tensions between strategic rhetoric and the ways of working used to deliver more resilient urban life. It is also required for the critical appraisal of vested interests that may yet absorb resilience into their own agendas, as true of sustainable development as of global capital. From a more theoretical standpoint such work may help significantly advance our understandings of the role of self-reflexivity and collaborative forms of responsible and responsive governance emerging within resilience-led governance practices and help us better understand the nature of creative disconnect between strategic policies and practical actions.

Bridging Strategy, Policy and Practice

This chapter set out to provide a backdrop for rethinking the interplay of urban growth, strategic policy and governance techniques being used to try and implement more resilient cities. The attempt to link strategy and policy to the practice of *resilient governance* treats resilience as a *process* not an outcome. In bridging strategy, policy and practice resilience is rendered manageable through a better understanding of the interactions between players across scales, not through the measurement of a metric by which the measure of resilience is improved – though the latter is far more amenable to the current strategic priorities of most interested parties. The above strategic frameworks seek to optimise resilience in a number of ways, including creating metrics, prioritising changes to expert service provision, demonstrating new developments as 'more resilient' than their predecessors and investing in local projects to test new ways of working. Of these the most progressive frame *resilience as a technique of governing* is inherently driven by its *methods*. More collaborative ways of working with a broader base of participants help to build resilience in ways difficult to measure but far more reflective of the rhetoric in strategy and policy. Strategies used to promote resilience building as a governance technique thus struggle to articulate how agencies can do their business in a different way whilst remaining cost effective, a key challenge to embedding resilience in urban governance *processes* rather than *resources*.

Resilience building may not always provide a *financial* dividend to investors, but rather it can improve the *conditions* within which interactions between stakeholders occur. Improving conditions for decision making which involves higher levels of public participation and collaboration leading to tangible outcomes (where the public input is visible) helps to build trust between experts and lay persons. This is a key capacity for working 'resiliently' – but again these are not core priorities in a system of governance dominated by economic growth. Resilience as a process may be read as an *institution*, for it plays out in the 'rules of the game'; in the process of being operationalised from strategy into techniques of problem solving. An analysis of implementation in different cities can help track how this is built into the fabric of urban governance. One cannot only focus on the physical components of the system or resources within the system, which are more malleable to linear forms of measurement; one must also include less linear, measurable and tangible forms of reflexive governance.

It is vital, if this indeed is the wider goal of a broader resilience project, that civil society is considered a key aspect of resilience. Civil society is the 'policy space' for the coordination of projects and the arbiter of exchanges between stakeholders. Creating resilience can be seen as a reinvigoration of civil society in and of itself, with the potential to mitigate the negative impact of half a century of neoliberal colonisation of urban governance for measurable economic growth. Concerns over this tension between neoliberal thinking and resilience thinking arise because there is a blurring of the edges between market-led governing at a distance and the progressive capacity-oriented logic embedded in the resilience strategies discussed here. Where responsibility for civil society functions is passed to non-governmental actors, understanding which organisations are dominant and how they work are thus vital to understanding what form of urban resilience emerges.

Resilience-led strategy may lay the conceptual groundwork for a shift in the core functions of organisations responsible for governing cities, but for this to take root third-sector organisations must be able to collaborate more closely with civil society (Otto, 1996; Teegen et al., 2004) and potentially to have direct access to local situations outside of market or legislative intervention when decisions on key issues of land use, planning, design, development and management roll out in practice. The 'top down' of global policy and the 'bottom up' of smaller projects *can* be bridged through a more nuanced reading of the interplay between strategic goals and the practices of governing. This must be developed further if 'where we may go from here' as presented in this chapter is to be useful in researching urban resilience.

A further hurdle comes from understanding resilient *capacities*. Capacities as presented here are perhaps not best understood as *outcomes created* by an investment in resilience governance, but as *conditions required* to enable more resilient outcomes to manifest from those investments. This is not a complete change to the 'cause and effect' model – i.e. investment leading to measurable outcome – lodged deep in neoliberalised governance practice, but rather a nudge of its trajectory towards allowing non-linear understandings of complex systemic interdependencies to be possible. For resilient governance one may thus surmise that it is no longer the job of experts/agencies to 'fix the problem', nor to 'fix the citizen' at risk, but rather to enhance the conditions within which all stakeholders are able to act in resilient ways. By allowing for 'resilience-led' self-determined actions *by all actors* negative impacts *may* be reduced, but if the professions aren't able to manage decisions for public interest over market-led efficiency metrics this is a fragile hope.

A key danger is the reduction of resilience to a metric or indices which once reached mean resilience can be forgotten about. Capacity enhancement when treated as a measurable outcome falls within a linear ontology of citizen–state relations; suggesting that citizens should be *shaped* by intervention. The implication of resilience thinking in policy rhetoric appears to imply one should treat citizens as 'members of heterogenous communities of allegiance' (Miller and Rose, 2008, page 25); quite different from a model of neurotic shareholders[3] in rational markets, where one can only shop for the most resilient option one can afford. A non-linear framing implies citizens are capable of acting in their own self-interest through a self-reflective form of agency. Individual actions manifest 'collective good' for both individuals and the wider community through reductions of both exposure and vulnerability, which offer an uncomfortable echo of the economic logic of non-equilibrium as a process for creative self-organisation in complex systems (Buchanan and Vanberg, 1991); but one with a value-driven (i.e. altruistic) understanding of the rational citizen-subject at its core.

End Game or Just Beginning?

The resilience that emerges through global policy platforms is likely to be aligned differently to the vagaries of each case and in each context within which it is encountered. This is a necessary feature of resilience projects that are implemented in such diverse situations. Equally, the capacities of actors will vary and so nuanced assessment techniques are required to identify vulnerabilities, a problem of which governance organisations are aware – as shown in the different approaches discussed previously. This is a key point. *There is no one final form of resilience done 'right'* regardless of a given project's success or of an organisation's influence or of an individual actor's opinion. What kind of resilience is being built, for what purpose, by whom and who benefits will, and must, vary in each case. The capacities approach to urban resilience in particular must be appreciated as 'enmeshed' within local contexts if it is to be fit for the purpose of the local people in each city where it is implemented. What works in Jakarta will likely not work in Tokyo. What works in Newcastle upon Tyne will likely not work in Medellin. This is good, so far as it allows for different versions of resilience to be mobilised dependent on the goals of the organisation driving the agenda wherever they may be. The organisation driving funded projects (often an NGO) have a huge role to play in appreciating what is possible given the local context of smaller-scale policy experiments. Such experiments are commonplace as a modus operandi of neoliberal governance strategy (Rose, 1999) and are often funded by NGOs as pilot or one-off projects under a broader public policy framework; thus seeking to increase the evidence base for a particular way of thinking or working in a format suitable to the evidence-based policy preferred by contemporary governments (Behague et al., 2009; Solesbury, 2001).[4] Evidence-based policy as a modus operandi of governing does however tend to encourage an uncritical 'scaling up' of successful experiments outside of the local context in which they were initially implemented, making a global movement aligned to a specific set of best practices or international standards very tricky,[5] if not contrary to the very logic of resilient governance itself.

One must also be careful not to be carried away by the potential of a transformative rhetoric. Any change to the methods, tactics and mechanisms of governing must be negotiated through existing operational processes embedded in organisations driven by decades of managerialism and neoliberalising project evaluation methods – often in terms of a rate of return on investment. Capacity building is not set up as a one-shot solution (throwing money and distant experts at the problem) but rather placing a little pressure on many disparate areas of action to enable individuals within those locales to have a broader set of choices available to them for solving problems locally, for citizens through their own agency and action, but also for practitioners enmeshed in their own organisational cultures. Such working may help to empower individuals by increasing access to more than services, but this variegated policy experimentation is itself vulnerable to resistance by experts locked into established workplace cultures, tactics and mechanisms of working; a tension strategic guidance fails to acknowledge.

As a governance strategy the implied potential is one of reducing both risk and dependency on external aid for vulnerable populations within their own 'micro-domains', wherein the true life of the public sphere is acted out – i.e. schools, church, sports clubs, etc. – as elements of nuanced community formation (Mulligan et al., 2016). This is usually operationalised and implemented by funding the launch of targeted project-funding schemes, however, it is also the stage where strategic rhetoric is subject to its widest interpretation by vested interests. Targeted subprojects are intended to focus on projects appropriate to local needs in which the citizen is the key stakeholder and driver, rather than the state. This gives greater weight to existing expertise in such areas where work has focused on local community schemes – such as flood warden schemes (Cabinet Office, 2008), participatory budgeting (UNESCO, 2016) and support for housing rights and land tenure in slum communities (Boonyabancha, 2011). But it is also where strategy meets reality and the gap between theory and implementable projects with measurable outcomes begin to take a firmer grip on the direction of actions taken on the ground.

It is useful to remain sceptical of project-led investment where a rate of return is assessed through linear managerially determined metrics, for fostering innovations in empowerment and increasing access to services may not be immediately measurable. Capacities-oriented resilience and resilient forms of governance are shifting interplays which manifest most clearly after a disaster; pre-emptive improvement in these capacities is thus iterative and hard to measure. This is true for both experts and for local people, and one must be on guard against top-down injections of 'first-world' solutions into local contexts for which they are inappropriate. A clear gap emerges between positive and negative potentials in translating resilience thinking from strategic frameworks into implementable projects, for the goal of experts nor government is not to abandon the individual to their own decisions (as the critics of 'neoliberalising' forms of resilience may suggest), but rather seeking to steer them to a broader range of options for self-determination by facilitating and/or coordinating the existing resilience-oriented capabilities of the community of practice, including both citizens as end users *and* practitioners in how they deliver resilience-led projects. Such capabilities have often been diminished or ignored by existing managerial forms of governance, by capital interests and by experts 'governing at a distance'. However, such *capabilities* and *capacities* are vital to building resilience in spite of, rather than to serve, economic interests or returns on investment.

Conclusions

This chapter is not designed to be a comprehensive or final word on urban resilience and global policy. It is designed to stimulate debate on the possibilities for thinking about urban governance in new ways. It is possible that the growth of urban resilience may present opportunities to expose the limitations of neoliberal governance of the urban growth machine in more detail. This could help critics to draw out the positive potential of global strategic platforms frameworks that seek to build resilience, but struggle to implement this in practice. Building a greater body of evidence for the unintended benefits of collaborative ways of working will help bolster the more progressive 'bottom-up' resilience projects, but there are no guarantees that this will lead to a step-change away from urban growth models. Adopting governance techniques that balance the privatisation of responsibility inherent in neoliberal governance with the empowerment at the heart of collaborative and participatory forms of community building and integrated urban development central to the resilience rhetoric is a long-term project affecting all of civil society. It is hoped that by identifying some connections between scale and scope – whereby resilience is filtered through international policy platforms into local urban decisions – readers can relate this appraisal of the challenges into their own research on the resilient city and, perhaps, begin to offer some tentative hypotheses for realising positive and progressive change in the management of the city as a 'resilience machine' producing something more lasting than short-term profits or new markets for exploitation. With the outcome unknown, we can, for now, assume that the interplay between urban growth, strategic frameworks and the diverse governance techniques adopted to implement more resilient cities will continue for some time to come.

Notes

1 See Rogers (2012) for more detail on praxis and poiesis in resilience research.
2 The subject of civil society will be returned to in greater detail later when reviewing the implications of global resilience, but more critical discussions can be found in the work of John Keane (1988).
3 A strong criticism of such tactics of governing can be explored through agency, reflexivity and risk taking or prudential behaviour (Walklate and Mythen, 2010).
4 An excellent critique of this trend can be found in Marston and Watts (2003).
5 Uvin addresses the tensions in scaling up and scaling down in his useful (1995) article.

References

Alexander, D.E. (2013). 'Resilience and disaster risk reduction: an etymological journey'. *Natural Hazards and Earth System Sciences*, 13(11): 2707–2716.
Attorney General (2011). *National strategy for disaster resilience*. Canberra: Attorney General's Department. Available at: www.ag.gov.au
Behague, D., Tawiah, C., Rosato, M., Some, T. and Morrison, J. (2009). 'Evidence-based policy-making: the implications of globally-applicable research for context-specific problem-solving in developing countries'. *Social Science and Medicine*, 69(10): 1539–1546.
Boonyabancha, S. (2011). *ACHR's work on housing rights over the last 23 years*. Bangkok. Available at: www.achr.net/
Buchanan, J.M. and Vanberg, V.J. (1991). 'The market as a creative process'. *Economics and philosophy*, 7(2): 167–186.
Cabinet Office (2008). 'Case study: community flood warden scheme'. Available at: www.gov.uk/government/publications/community-resilience-case-study-library

Christopher, M. and Peck, H. (2004). 'Building the resilient supply chain'. *International Journal of Logistics Management*, 15(2): 1–14.

Civil Contingencies Secretariat (2004). *Civil Contingencies Act 2004: a short guide (revised)*. London: Cabinet Office.

Coaffee, J., Wood, D.M. and Rogers, P. (2009). *The everyday resilience of the city*. Basingstoke: Palgrave Macmillan.

Cutter, S.L., Barnes, L., Berry, M., Burton, C., Evans, E., Tate, E. and Webb, J. (2008). 'A place-based model for understanding community resilience to natural disasters'. *Global environmental change*, 18(4): 598–606.

Filion, P. (2013). 'Fading resilience? Creative destruction, neoliberalism and mounting risks'. *SAPIENS: Surveys and Perspectives Integrating Environment and Society*, 6(1).

Godschalk, D.R. (2003). 'Urban hazard mitigation: creating resilient cities'. *Natural hazards review*, 4(3): 136–143.

HABITAT III (2015). *Issue Papers 15: Urban resilience*. New York. Available at: http://unhabitat.org/issue-papers-and-policy-units/

Handmer, J.W. and Dovers, S.R. (1996). 'A typology of resilience: rethinking institutions for sustainable development'. *Organization and Environment*, 9(4): 482–511.

Jay, M. (1984). *Adorno*. Cambridge, MA: Harvard University Press.

Joseph, J. (2013). 'Resilience as embedded neoliberalism: a governmentality approach'. *Resilience*, 1(1): 38–52.

Keane, J. (1988). *Civil society and the state: new European perspectives*. London: Verso.

Klein, R.J., Nicholls, R.J. and Thomalla, F. (2003). 'Resilience to natural hazards: how useful is this concept?' *Global Environmental Change Part B: Environmental Hazards*, 5(1): 35–45.

Lee, A.V., Vargo, J. and Seville, E. (2013). 'Developing a tool to measure and compare organizations' resilience'. *Natural Hazards Review*, 14(1): 29–41.

Magis, K. (2010). 'Community resilience: an indicator of social sustainability'. *Society and Natural Resources*, 23(5): 401–416.

Marston, G. and Watts, R. (2003). 'Tampering with the evidence: a critical appraisal of evidence-based policy-making'. *Drawing Board: An Australian Review of Public Affairs*, 3 (3): 143–163.

Miller, P. and Rose, N. (2008). *Governing the present: administering economic, social and personal life*. Cambridge: Polity.

Mulligan, M., Steele, W., Rickards, L. and Fünfgeld, H. (2016). 'Keywords in planning: what do we mean by "community resilience"?' *International Planning Studies*: 1–14.

Otto, D. (1996). 'Nongovernmental organizations in the United Nations system: the emerging role of international civil society'. *Human Rights Quarterly*, 18(1): 107–141.

Pelling, M. and Dill, K. (2010). 'Disaster politics: tipping points for change in the adaptation of sociopolitical regimes'. *Progress in Human Geography*, 34(1): 21–37.

Ponomarov, S.Y. and Holcomb, M.C. (2009). 'Understanding the concept of supply chain resilience'. *International Journal of Logistics Management*, 20(1): 124–143.

Reghezza-Zitt, M., Rufat, S., Djament-Tran, G., Le Blanc, A. and Lhomme, S. (2012). 'What resilience is not: uses and abuses'. *Cybergeo: European Journal of Geography*.

Resilience Alliance (2016). *Resilience Alliance*. Available at: www.resalliance.org/

Rockefeller Foundation (2016a). *100 Resilient Cities*. Available at: www.100resilientcities.org/

Rockefeller Foundation (2016b). *Resilience city strategies*. Available at: www.100resilientcities.org/strategies#/-_/

Rogers, P. (2011). 'Resilience and civil contingencies: tensions in northeast and northwest UK (2000–2008)'. *Journal of Policing, Intelligence and Counter Terrorism*, 6(2): 91–107.

Rogers, P. (2012). *Resilience and the city: change, (dis)order and disaster*. London: Ashgate Publishing.
Rogers, P. (2015a). 'Researching resilience: an agenda for change'. *Resilience*, 3(1): 55–71.
Rogers, P. (2015b). 'Resilience as standard: risks, hazards and threats', in T. Balzacq (ed.), *Contesting security: strategies and logics*. London: Routledge, pp. 189–204.
Rogers, P. (2017). 'The etymology and genealogy of a contested concept', in D. Chandler and J. Coaffee (eds), in *The Routledge Handbook of International Resilience*. London: Routledge, pp. 13–25.
Rose, N. (1999). *Powers of freedom: reframing political thought*. Cambridge: Cambridge University Press.
Solesbury, W. (2001). *Evidence based policy: whence it came and where it's going*. London: ESRC UK Centre for Evidence Based Policy and Practice.
Stockholm Resilience Centre (2016). *Stockholm Resilience Centre*. Available at: www.stockholmresilience.org/
Teegen, H., Doh, J.P. and Vachani, S. (2004). 'The importance of nongovernmental organizations (NGOs) in global governance and value creation: an international business research agenda'. *Journal of International Business Studies*, 35(6): 463–483.
Thomalla, F., Downing, T., Spanger-Siegfried, E., Han, G. and Rockström, J. (2006). 'Reducing hazard vulnerability: towards a common approach between disaster risk reduction and climate adaptation'. *Disasters*, 30(1): 39–48.
Tierney, K. (2015). 'Resilience and the neoliberal project: discourses, critiques, practices – and Katrina'. *American Behavioral Scientist*, 59(10): 1327–1342.
Timms, B.F. (2011). 'The (mis)use of disaster as opportunity: coerced relocation from Celaque National Park, Honduras'. *Antipode*, 43(4): 1357–1379.
UK Resilience (2005). *Emergency preparedness*. 23 November.
UNESCO (2016). *The experience of the participative budget in Porto Alegre Brazil*. Available at: www.unesco.org/most/southa13.htm
United Nations (2015a). *Sustainable development knowledge platform*. Available at: https://sustainabledevelopment.un.org/sdgs
United Nations (2015b). *Transforming our world: the 2030 agenda for sustainable development*. Available at: www.un.org/ga/search/view_doc.asp?symbol=A/RES/70/1&Lang=E
United Nations (2015c). *The future we want: Resolution 66/288*. Available at: www.un.org/ga/search/view_doc.asp?symbol=A/RES/66/288&Lang=E
United Nations (2016). *Making cities resilient: my city is getting ready*. Available at: www.unisdr.org/campaign/resilientcities/
Uvin, P. (1995). 'Scaling up the grass roots and scaling down the summit: the relations between Third World nongovernmental organisations and the United Nations'. *Third World Quarterly*, 16(3): 495–512.
Walklate, S. and Mythen, G. (2010). 'Agency, reflexivity and risk: cosmopolitan, neurotic or prudential citizen?', *British Journal of Sociology*, 61(1): 45–62.
World Economic Forum (2013). *Building resilience in supply chains*. Geneva. Available at: www.weforum.org/

9
TOWARDS A CRITICAL POLITICAL GEOGRAPHY OF RESILIENCE MACHINES IN URBAN PLANNING

Thilo Lang

> There is need to theoretically emancipate resilience as a concept, sensitive to its historiography. As a heuristic this plurality of resilience theories are very useful tools. They allow us to simplify extreme complexity. However, the way resilience is deployed in academic and political spheres seems to suggest that that simplification might itself be mistaken for the messy materiality of life in all its forms and species ... The inherent danger is that policy and academic analysis becomes concerned with understanding and maintaining a system shorn of political context or attention to questions of power and inequality.
>
> *(Welsh, 2014, page 21)*

Introduction

In the past years, resilience has turned into one of the major buzzwords in urban studies as well as in urban politics. This coincides with seemingly diverse and expanding challenges cities across the world have been exposed to in the past years (such as the simultaneity of old and new economies, their function as entry points for migrants and refugees, various sorts of perceived security threats, climate change, growing urban–rural polarisation, issues of exclusion and marginalisation, urban protest movements, increasing global competition and so on). Increasingly, policy makers and urban planners see the promotion of resilience as a promising way forward at times when it seems to be more and more difficult to respond to challenges impossible to predict. For many, resilience is the key to the governance of complexity (see Chandler, 2014a). But, to what extent is the notion of urban resilience a useful approach in this context? And what understanding of resilience should we follow?

Based on a review of the current relation of urban planning to these new forms of uncertainty, this chapter discusses the limits and opportunities of the resilience machine out of a critical urban and regional studies perspective. The argument that grounds this chapter is that work on resilience often is based on the normative idea that if we better understand how the underlying systems work, we would still be able to control the system and plan for better resilience. As this seems slightly paradox, mechanistic understandings of resilience disclosed, and resilience should better be conceptualised in the light of assemblage theory implying 'that the resilience machine is produced through the ensemble of actual and potential relations between its heterogeneous components' (Davoudi et al., this volume). Further, we should differentiate resilience as a normative political concept from alternative analytical instruments to research adaptive capacities in the context of the resilience machine.[1] Such an approach would highlight the institutional preconditions for dealing with rising complexities, multiplying rationalities and non-controllability in and of urban development.

The remainder of this chapter is structured as follows. First, the role of growing uncertainties, multiplying rationalities and rising complexity of to-date post-modern societies for the rise of debates on urban resilience will be discussed and the idea of adaptive capacity linked to the resilience perspective explored. Based on an analysis of the more recent critical literature on resilience, this chapter will continue to challenge some dominant logics of promoting resilience and ask for a critical political geography approach linked to evolutionary and complexity thinking. This chapter concludes with five propositions for a more critical grounding of research on the resilience machine.

Setting the Scene: Urban Planning in Times of Uncertainty and Rising Complexity

For a long time – at least in the 'Western world' – in urban and regional planning, the dominant thought was that the future is open and can be modified and shaped through intelligent planning (Taylor, 1998). More recently, the deeply internalised logics of rationality in urban planning has been called into question and a paradigm shift seems under way towards a worldview based on a different ontology (e.g. Davoudi et al., 2012, page 302). What if we conceptualise urban development as the result of unexpected events and side effects of purposeful action? What if we perceive urban development as non-controllable – as a wicked problem? Our understanding of the role of planning and the scope for public policy (e.g. in economic development promotion) might have to be redefined.

In recent decades, some scholars challenged the idea of rationality in planning theory, responding to growing complexity which has been recognised as a major feature of post-modernity (Portugali, 2011; de Roo and Silva, 2010). Since the initial critiques on the rationality of planning in the late 1960s, many approaches were directed towards simplifying complexity and understanding structures shaping complex systems in order to achieve a higher degree of certainty and to enhance the scope for planning (e.g. Koch et al., 2015, page 11). At the same time, planning theory started to focus on the 'making' of places and 'better' governance, coordination of interests and mobilising power (Healey, 2010). Urban planning has concentrated on inventing 'new ways for gathering resources and organizing actors to act together in a synchronized way on the urban realm' (Jihad and Jacques, 2012, page 95, for a critical take on this focus on governance, see e.g. Jessop, 1997).

Despite these shifts, planning theory and practice have not fully managed to escape the 'traumata' of the loss of controllability. In particular across Europe, comparative planning systems research has revealed traditional modernist legacies of disciplinary training that perceive planning as a structured rational process oriented towards the future – reflecting the general order of the nation state operating under functionalist and structuralist logics and methodologies (Reimer et al., 2014).[2]

Complexity and uncertainty related to urban development can be seen as marking the antonyms of rationality and intentionality of urban planning. In the past years, planning theory has oscillated between conceptualisations linked to these two extremes, maintaining the idea that urban development should (and can) be planned for and controlled. Although the initial critique on the rationality of urban planning was brought forward more than 40 years ago: 'planning practice is still much the same as it ever was … this might have something to do with the fact that time and time again … alternatives were still formulated within the existing planning framework' (Boelens, 2010, page 30). It is in that sense not surprising that different approaches to urban planning evolved from outside of the academic planning community, whereas planning theory has dealt with issues of complexity and uncertainty mainly in a normative way (developing approaches how planning should be to be better able to coordinate rising complexities). To achieve a different understanding of the current challenges of professionals in urban planning and policy, this chapter draws on some classic sources of early critique and highlights factors which so far have been under-researched, such as the role of unexpected events, coincidence and chance, side effects and unintended consequences of action – urban planning as a wicked problem (Rittel and Webber 1973), windows of opportunity (Boschma 1996; Olsson et al., 2006) and a garbage can model of organisational choice (Cohen et al., 1972):

- As a trained mathematician and physician, Horst Rittel (Rittel and Webber, 1973) proposed as early as 1969 that in a pluralistic society many problems of social policy are too complex to be solved with standard recipes based on rational choice. Moreover, he argued that science generally has developed to deal with simple problems whereas policy problems usually would escape clear description. Rittel thus coined the term 'wicked problems'.

- Whereas planning professionals and local politicians often relate 'successes' in urban development directly with their work, evolutionary economist Ron Boschma (1997) argued that in particular for new industries, it is impossible to predict where they would evolve because location decisions would be determined to a large extent by factors that it is not possible to predict and outside of the scope of local action. Olsson et al. (2006) stress that also in the sphere of policy and governance, the coincidence of identified problems, available solutions and political will are decisive to achieve change.
- A third approach stems from organisation theory and management. Cohen et al. (1972) turned the decision-making logics in organisations around. Their basic presumption was that large organisations in particular would always be characterised by some form of anarchy. Decisions in organisational contexts would rather be based on available solutions (than on actual knowledge), upcoming problems, decision opportunities and the more or less random combination of participants. They would therefore escape rational legitimisation.

Taking these three perspectives together, urban development should be perceived as to a large extent escaping rational control. It is in this context that notions of resilience and complexity gained importance in urban planning as both appear to provide solutions to managing rising challenges cities and political actors are confronted with. For many actors, supporting resilience might come with the promise to offer a one-size-fits-all solution to a range of issues, including those impossible to predict. Hence, at a superficial level, promoting resilience risks creating the illusion that managing urban development is still possible imposing a new rationality. For more critical scholars, resilience appears as a new form of governing complexity (Chandler, 2014a). The key here is to accept that it is difficult to simplify complexity and that planning will have to shift its focus, acknowledging new forms of governance in line with these complexities. A form of governance that accepts wicked problems and that coincidence, side effects and anarchic forms of decision making are usual components of urban development, opposing the belief in rational system-controllability.[3]

Complexity, Urban Planning and Dealing with Nescience

The idea that in principle all issues are subject to rational control is seen as a common feature of modern sciences (Habermas, 1968; Kreibich, 1986). However, under the conditions of multiplicity, scientific abstraction, controllability and predictability are more and more prone to failure, and deviations from envisaged outcomes through planned processes as well as uncertainties have turned into normality (Böhle et al., 2001, page 97). A number of scholars in complexity and system theory (such as Walby, 2007; Corning, 2002; Byrne, 1998; Ratter, 2012; Luhmann, 1984) have worked on issues of emergence in order to understand outcomes of social processes based on the unstructured and complex intersection of non-homogeneous rationalities of numerous agents following individual agendas. As theories of complexity usually stress that the whole is more than the sum of its components, emergent properties, structures and behaviour are unpredictable. The potential number of interactions between components of social systems increases exponentially with the growing complexity of a system. It is also due to these interactions that unintended consequences and side effects play a major role in complex systems.

Beck et al. (2001a, 2003) offer a kind of metaframework to better understand the rising challenges of urban planning through their conceptualisation of unintended consequences, side effects, the multiplying of rationalities and the relation between knowledge (Wissen) and nescience (Nichtwissen) (see also Beck, 1996). The main idea of their theory of reflexive modernisation is that the so far taken-for-granted basic social principles and clear distinctions (between nature and society, between knowledge and belief, between expert and lay knowledge, between members of society and outsiders, between the nation state and the global or between economy and state) of modern society are now seen as being in a state of flux: 'the system of coordinates is changing' and 'the old certainties, distinctions and dichotomies are fading away' (Beck et al., 2003, page 2). The key question here is, 'how can one make reasonable decisions about the future under conditions of such uncertainty?' (page 3).

How (and for whom, and with which consequences) could it be made possible to communicate nescience? Would it be possible to take political decisions on the basis of nescience (Beck et al., 2001b, page 77)? In a controversy with Anthony Giddens and Scott Lash, Beck examines five dimensions of nescience: selective reception and communication, uncertain knowledge, errors and mistakes, the impossibility to know (which again can be known or ignored) and the unwillingness to know (Beck, 1996, page 302). We can use these dimensions to work empirically on practices of planning and political decision making under conditions of uncertainty based on different dimensions of nescience and knowledge. Beck highlights here that we should not conceptualise nescience as potential knowledge (things that we do not yet know) and should rather focus on things impossible to know (Beck, 1996, page 304 and following pages) or the 'unknown unknowns' as Chandler (2014b, page 50) puts it. In the context of the resilience discourse, it is in particular this form of nescience which is relevant in its ontology describing events, outcomes and causalities impossible to know beforehand.

Urban Planning, System Thinking and Resilience in the Context of Rising Complexities

Although there is growing recognition of complexity and rising uncertainties (Luhmann, 1984, 1997; Böhle and Busch, 2012; Corning, 2002; Manson, 2001; Davoudi et al., 2012) playing a major role in urban development, and although some planning theorists have been relating to post-structural and post-modern conceptualisations of space and place (Davoudi and Strange, 2009), different approaches to planning 'had the difficulty to assert themselves and conquer the planning practice' (Jihad and Jacques, 2012, page 94). Further to that, many of the new terms linked to these approaches carry connotations which appear to be negative: 'muddling through', 'bricolage planning', 'incremental planning', 'planning as experimentation and speculation' and so on. Such terms do not sound convincing in providing a theoretical orientation to planning practice. The basic insight from these readings might be that planning theory and practice need to shift towards a pragmatic approach overcoming unrealistic expectations of controllability (see also Hillier, 2010, page 448; de Roo, 2010, page 15).

Resilience thinking comes with a positive language focusing on networking, resources and adaptive capacities: 'Increasingly it has been recognized that the key to overcoming such difficulties is a better engagement with key stakeholders, that is, those who do exercise relevant resources and/or hold relevant information and knowledge' (Rydin, 2010, page 266). However, there is no common understanding of the notion of urban resilience and academic as well as political debates often seem a little fuzzy and blurred. The only basic agreement across most writing on resilience is system thinking (whereas there are great variations in how to understand and delimit the system). Another common position is to apply a kind of equilibrium model linking resilience to the 'ability' of a city or region to return to its earlier, pre-crisis state or to be transformed into a new stable version of the system (whereas there is rising criticism on equilibrium thinking and bounce-back capacities).

Many reviews on resilience start with noting the absence of a common definition of the term and its conceptual fuzziness. This most likely relates to the diverse range of ontological and epistemological perspectives utilised and the disciplinarily cross-cutting appeal of the term. Leaving aside approaches referred to as engineering resilience based on equilibrium thinking and resilience approaches linked to disaster management, understandings derived from socio-ecological research based on complexity thinking are offering an alternative (see also Chandler, 2014a and 2014b). Such understandings link up to resilience research which is rooted in post-positivistic epistemologies (contrast Walker et al., 2006), applying an understanding of cities and regions as complex adaptive systems (following Holling, 1973). Whereas concepts of resilience are often used to describe the relationship between the system under observation and the externally induced disruption, stress, disturbance or crisis, some scholars stress evolutionary resilience as the capacity for constant adaptation, change or transformation (compare Davoudi et al., 2012, 302 and following pages). Thus, resilience is more than a response to or the means of coping with particular challenges. Resilience could be seen as a kind of systemic property. Although thinking about urban resilience, in the past decades, has borrowed from other disciplines, the major theoretical body exploited for a conceptualisation of urban and regional resilience stems from socio-ecological research. In order to develop a more qualitative take on urban and regional resilience based on complexity thinking, it is worthwhile to review the conceptualisation of adaptation in the context of socio-ecological systems.

For Holling (2001), the understanding of socio-ecological resilience is closely related to questions of adaptation. He describes complex adaptive systems of people and nature as being self-organised, with a few critical processes creating and maintaining this self-organisation. Such systems of nature, humans, combined human-natural systems, or social-ecological systems can be perceived as being 'interlinked in never-ending adaptive cycles of growth, accumulation, restructuring, and renewal' (page 392). The accumulation and transformation of resources alternate with phases creating opportunities for innovation. Understanding these cycles, their temporal and spatial scales, as well as the relevant frames of reference used to produce and value these processes would help to 'identify the points at which a system is capable of accepting positive change and the points where it is vulnerable' (page 392). In this context, (a) resilience as 'a measure of its [the system's] vulnerability to unexpected or unpredictable shocks' (page 392), (b) the internal controllability and (c) the wealth of the system determining the range of possible future options are seen as the main properties shaping the adaptive cycles and the future state of the systems (page 393). High resilience would allow for tests of novel combinations that trigger innovation and adaptation. Holling sees this as particularly true if controllability is low and high resilience allows for the recombination of elements in the system because the costs of failure are low. In contrast, with low resilience or vulnerability at multiple scales, revolutionary transformations would be quite rare due to the nested character of sets of adaptive cycles. A combination of separate developments would have to coincide, i.e. there would be a need for recombinations and inventions to occur simultaneously in order to open windows for fundamental new opportunities (compare Holling, 2001, page 404).

In offering an initial conceptualisation of adaptation, it is not surprising that Holling's ideas have been transferred from the study of social-ecological systems to the general sphere of urban and regional development. Hence, in urban and regional studies, adaptive systems can be seen as a set of systemic properties to anticipate or address problems and change or transform in a way that generates long-term stable development paths and allows cities and regions to adapt through external interference. Finally, resilience should be perceived as a dynamic process instead of a static state: 'Resilience in this perspective is understood not as a fixed asset, but as a continually changing process; not as a being but as a becoming' (Davoudi et al., 2012, page 304).

Towards a Critical Political Geography of Resilience Machines

If we perceive resilience as a systemic property to cities and regions in the sense of adaptive capacities, how should we define the system, and to what extent can we support the emergence of resilience? Is there any analytical value to the notion of resilience? And to what extent can empirical research inform policy makers about the potential value for resilience as a normative political concept?

Applying a Scalar and Relational Approach to Cities and Regions

Who is 'the city' if talking about resilient cities? As with an ecosystem, urban and regional socio-economic systems cannot be seen as working within given boundaries and their multiscale interdependencies have to be acknowledged along with their complex specific institutional environments (Lang, 2009, page 194 and following pages). Such institutional environments include, besides particular local institutional frameworks, the interplay with national forms of regulation, the wider policy context and the (local interpretation of) macroregional and global framework conditions. When it comes to understanding variations in how cities and regions relate to external stress, we should include an analysis of structural and framework conditions or portray the ways decision takers and firms have dealt with external economic and social disturbances in the context of such conditions. For example Simmie and Martin (2010, page 42) conclude, after analysing governance and firm responses over a number of decades following an evolutionary logic of adaptive cycles, that the adaptive cycle model is a useful approach to analyse and better understand regional (economic) resilience.

In normative political contexts in particular, some scholars tend to seek specific local properties that make urban areas less vulnerable to (perceived forms of) crises, sometimes raising expectations about self-contained urban systems functioning like a perpetual motion machine. There is a risk in limiting research on urban and regional change and adaptation to internal regional properties. This ignores the crucial question of national regulation and the necessity to investigate the role of national frameworks and forms of intervention as well as political economic issues of globalisation. In times of increasing globalisation during contemporary capitalism, cities and regions are increasingly networked across the world by means of an internationalised economy with functional specialisation, global value chains and a spatial division of labour. Thus it becomes more and more impossible to understand how a particular urban or regional economy works based on what happens within particular administrative boundaries. We have to avoid the pitfalls of applying a container logic to cities and regions and ignoring these multiscalar relations. Due to the principles of how the economy currently functions, the system would have to be defined as a globally networked multilevel political economic system. Within this system, some places with particular forms of economic activities seem to be privileged while others are disadvantaged. Therefore, in most cases, the very reasons for urban decline and for a growing number of non-resilient cities and regions cannot be explained without referring to the ways in which present-day capitalism functions.

Take Institutional Dimensions and Power Relations Seriously

If we are using the term resilience machine, do we imply a kind of powerful resilience coalition promoting and implementing resilience thinking? Who are the drivers of local, regional or national resilience debates? Realising the normative overload in much of the current research on resilience, we should be sensitive to the inherent normative and paradigmatic values shaping these approaches. Perceived forms of crises and vulnerability as well as desirable ends are always socially constructed. In resilience research, questions tend to be overlooked as to why and how external events are perceived as disturbance or crisis, as well as questions on the achievable state after crisis. In this context, questions of power concerning the judgement and classification of what constitutes a crisis situation, as well as the objectives for achievable states of the system through adaptation are crucial and should be topics for further research (see also Hudson, 2010, page 13), as should issues regarding the social construction of vulnerabilities and resilience in general (see also Bürkner, 2010, page 34 and following pages.). In this context, the adaptation of a system does not have to correspond to widely accepted 'positive' developmental trajectories and would also allow for alternative futures. Future resilience research will have to avoid the risk of maintaining a system shorn of political context and questions of power and inequality.

In addition, systemic thinking should include the analysis of structural forces. Hill et al. (2008, page 2) stressed that such a perspective helps to shift the focus to analysing the long-term structure of macroeconomic relationships and the relevant social, economic and political institutions conditioning these structures. Hence, studying resilience should involve studying the rise, stability and decay of institutions conditioning long-term economic success within a governance dimension. This perspective might help towards a better understanding of why some cities maintain their development paths during various forms of crises and how they adapt to changing framework conditions induced by disruption, stress or crises. However, such a conceptualisation seems to be less suited to the understanding of persisting vulnerabilities.

Avoid Equilibrium Thinking and Allow for Alternative Conceptualisations of Development

Some scholars favour a perspective of resilience linked to the ability of cities and regions to bounce back or reachieve an equilibrium. But, what should be the 'normality' or the 'original state' the system should bounce back to? And wouldn't such a view discharge the transformative and adaptive capabilities of a system (e.g. Martin and Sunley, 2015) and incremental change (Gong and Hassink, 2017)? Instead of focusing on equilibrium, we should acknowledge the possibility of fundamental change. 'Recognising the capacity of systems to change, and the significance of "system resilience" as both a constrainer and enabler of alternative regimes' (Welsh, 2014), resilience can inform analytical frameworks to examine change itself.

I suggest to seek conceptualisations of resilience going beyond an engineering approach which perceives risk and disturbance as measurable and manageable (see also Beilin and Wilkinson, 2015). Instead, complexity theory would lead us to take the future as unpredictable and uncertain and resilience as the capacity to deal with this uncertainty. 'Emergent systems have uneven consequences that reflect both slower and faster variables than can always be anticipated' (Beilin and Wilkinson, 2015, page 1207).

Hudson astutely discusses options for more resilient regions in relation to the resilience of capitalism and dominant neoliberal models of regional development which trigger national state interventions, and which have at best been partially and temporarily successful in the past (Hudson, 2010). Indeed, neoliberal thought as a dominant feature of current capitalism can be seen as having become maladaptive and a major threat to urban and regional resilience. In particular in the fate of the financial, economic and national debt crisis, many countries have introduced austerity policies and measures prioritising the rescue of the financial systems without giving too much thought to the social consequences. These measures are usually justified by pointing out that there is no alternative and no choice. Hence, instead of working against the initial reasons, such as the deregulation of financial markets, such neoliberal policies are even strengthened, whereas cuts in the social sphere are accepted as unavoidable (see also Krätke, 2014).

Be Sensitive to Forms of Neoliberal Instrumentalisation

Opposing such radical forms of response, political ideas of promoting resilience as a normative model match currently dominant neoliberal policy agendas very well. In order to maintain the wider system, responsibility for critical developments risks being projected on an urban or regional level, combined with (implicit or explicit) request that such places should work on becoming more resilient. Hence, resilience thinking may well be exploited to reproduce contemporary capitalism instead of acknowledging the opportunities for transformation of the wider system logics. 'Resilient spaces are exactly what capitalism needs – spaces that are periodically reinvented to meet the changing demands of capital accumulation' (MacKinnon and Derickson, 2013, page 254). Thinking about resilience as a normative, externally defined strategy would even reproduce uneven forms of development, or at least would serve to reproduce 'wider social and spatial relations that generate turbulence and inequality' (page 254). However, such an instrumentalisation of the concept as a normative model contradicts the understanding of adaptive systems as being self-organised. Hence, raising adaptive capacities stands against political strategies to deresponsibilise risk away from the state: 'These resilience discourses are situated in, and help reproduce, broader neoliberal practices of security that shift from state based to society-based conceptions of distributed risk and reaction' (Welsh, 2014, page 19) and thereby maintain a status quo (see also Beilin and Wilkinson, 2015, page 1206). Hence, the idea of self-organisation linked to the system perspective should not be mistaken as self-reliance. 'Resilience machines' could also appear as powerful carriers of neoliberal policies directing the responsibilities for all sorts of risks away from the state and thereby reproducing existing forms of inequality and socio-economic polarisation. However, the misuse of the resilience concept in neoliberal ideology should not distract attention from progressive bottom-up processes (Mykhnenko, 2016). There is a plethora of opposition groups, grassroots campaigns, ecologist and idealist movements, amongst others, proclaiming the concept for their aims (as mentioned for example by MacKinnon and Derickson, 2013 or by Mykhnenko, 2016; for a concrete example see Goldstein et al., 2015).

Explore the Progressive Potentials of the Resilience Machine in Pursuit of Justice and Democracy

The main limitation of the resilience concept concerning the latter points is its being apolitical or depoliticised and thus lacking explanatory strength. A further hope projected onto the emerging resilience debate is that it provides us with a new approach for better understanding change and adaptation. Only by highlighting issues of (progressive) change and adaptation, the transformative potentials of the resilience concept can play out. Holling conceptualises the relation between resilience and adaptation with the panarchy model which nicely resonates with how we understand the concept of assemblage (see also Davoudi et al., this volume). In the panarchy model, adaptive cycles are portrayed as being never-ending, nested and complex, supporting the emergence of new structures and behaviours and allowing for novel combinations which potentially lead to innovation because the costs of failure are low. However, the approach fails to give an idea of how this might function in a social world and is criticised by some for being over-deterministic (see also Davoudi et al., this volume). This may be in particular because it lacks a conceptualisation of agency within the system perspective, including questions of understanding collective agency, power and agenda setting in a world of diverse interests. Therefore, studies using the concept would need to 'move beyond emphasis upon narrow … metrics of regional macro-economic performance … pre- or post-shocks' (Bristow and Healy, 2014, page 932). Otherwise results will mainly portray in a descriptive form the ways in which cities and regions cope with external stress from a purely economic point of view, while neglecting questions of long-term social stability. As the most recognised proponents of a complexity perspective on resilience, members of the Resilience Alliance themselves have added four key factors being decisive to prepare systems to change: building knowledge, networking, leadership and timing/windows of opportunity (Olsson et al., 2006). Joining all these positions leads to a conceptualisation of adaptive cycles being at the same time constraining and enabling. Resilience then can be seen as the capacity to identify windows of opportunity at least for gradual transformation while disruptive change would be less likely to occur due to the nested character of adaptive cycles.

Conclusions

In the past years, the concept of urban and regional resilience has entered both the political and the academic world. While a number of think tanks and governments at various scales and in various countries around the world started to promote resilience as a normative concept, many researchers have applied analytical readings of the resilience concept with a strong focus on mechanistic equilibrium thinking. This chapter argued for a different conceptualisation stressing the idea of adaptive cycles and the panarchy model as a form of assemblage. With taking this understanding seriously and inserting agency and power to the conceptualisation of adaptive cycles and resilience, we can avoid the risk that researching resilience becomes a nice exercise for economists in sorting regions according to economic indicators without going deeper into the issues of uneven development. Further, we can better understand the progressive potentials of the concept stressing its transformative power.

Following the five propositions to respect when developing urban research agendas based on resilience thinking, we should next to widespread criticism also respect the merits of resilience as potentially transcending disciplinary boundaries and offering a new approach to theorise urban and regional transformation suggesting a focus on (redundant) institutional capacities to better deal with external shocks and disturbance (see also Mykhnenko, 2016, 202 and following pages). In this way, resilience thinking also offers a framework to include coincidence, windows of opportunity and wicked problems in a conceptualisation of urban and regional transformation which goes beyond our current knowledge about institutional change, but still regards institutions as being components of the wider system.

As a final contribution to this chapter, two major avenues for further research in the field can be identified. First, there is an urgent need to better understand the underlying rationales of political agendas and think tank activities in promoting resilience. What is the ultimate purpose of these initiatives and what are their wider implications for individuals, communities and local administrations in this context? Further elaborating on the notion of 'resilience machines' has the potential to propose a research agenda focusing on issues of (political) power acknowledging the wider reasons for uneven development. This would further help to overcome the conceptual fuzziness of current research on urban resilience. Working on resilience machines would allow researchers to explore the meaning and utilisation of the concept as applied in political debates and in official rankings generated to facilitate regional policy. Second, there is great potential merit in further conceptualising the progressive potential of resilience as a capacity to mitigate positive change and transformation. How can we better understand and support adaptive change and the opening of new paths of development when previous ones fail to distribute wealth to all urban communities? Thinking about urban development inspired by an assemblage perspective on resilience and adaptive cycles would also allow urban and regional planners to reinvent their scope for action beyond their traditional understanding as rational and technical professionals. This would open up alternative ways for adaptive management and a 'postmodern form of governance' (Chandler, 2014b, page 47), embracing the complexities of our contemporary world.

Notes

1 While acknowledging the need for further research about how the concept is used in normative contexts, I sympathise with advancing the analytical dimension of the resilience concept linked to panarchy and adaptive cycles.
2 The current euphoria around 'big data' can be read as a major backlash as it seems to reflect the idea that with big data one can better predict and anticipate development and (re)achieve the capacity for 'better' planning (French et al., 2015; Batty, 2013; Anderson, 2008).
3 This does not mean that there is no longer any need for rational planning. Indeed, some subfields of planning (such as disaster management) do need a lot of conventional and rational planning.

References

Anderson, C. (2008). 'The end of theory: the data deluge makes the scientific method obsolete'. *Wired Magazine*, 16(7). Available at: www.wired.com/2008/06/pb-theory/
Batty, M. (2013). 'Big data, smart cities and city planning'. *Dialogues in Human Geography*, 3 (3): 274–279.
Beck, U. (1996). 'Wissen oder Nicht-Wissen? Zwei Perspektiven "reflexiver Modernisierung"', in U. Beck; A. Giddens, A. and S. Lash, S. (eds), *Reflexive modernisierung. Eine kontroverse*. Frankfurt a.M.: Springer, pp. 289–315.
Beck, U., Bonß, W. and Lau, C. (2001a). 'Theorie reflexiver Modernisierung - Fragestellungen, Hypothesen, Forschungsprogramme', in U. Beck and W. Bonß (eds), *Die Modernisierung der Moderne*. Frankfurt a.M.: Suhrkamp Verlag, pp. 11–62.
Beck, U., Holzer, B. and Kieserling, A. (2001b). 'Nebenfolgen als Problem soziologischer Theoriebildung', in U. Beck and W. Bonß (eds), *Die Modernisierung der Moderne*. Frankfurt a.M.: Suhrkamp Verlag, pp. 63–81.
Beck, U., Bonß, W. and Lau, C. (2003). 'The theory of reflexive modernization: problematic, hypotheses and research programme'. *Theory, Culture and Society*, 20(2): 1–33.
Beilin, R. and Wilkinson, C. (2015). 'Introduction: governing for urban resilience'. *Urban Studies*, 52(7): 1205–1217.
Boelens, L. (2010). 'Theorizing practice and practising theory: outlines for an actor-relational-approach in planning'. *Planning Theory*, 9(1): 28–62.
Böhle, F. and Busch, S. (2012). *Management von Ungewissheit: Neue Ansätze jenseits von Kontrolle und Ohnmacht*. Bielefeld: Transcript.
Böhle, F., Bolte, A., Drexel, I. and Weishaupt, S. (2001). 'Grenzen wissenschaftlich-technischer Rationalität und "anderes Wissen"', in U. Beck and W. Bonß (eds), *Die Modernisierung der Moderne*. Frankfurt a.M.: Suhrkamp Verlag, pp. 96–105.
Boschma, R.A. (1996). 'The window of locational opportunity – concept'. *Collana di Teoria Economicà*, 43(260): 1–36.
Boschma, R.A. (1997). 'New industries and windows of locational opportunity: a long-term analysis of Belgium'. *Erdkunde*, 51: 12–22.
Bristow, G. and Healy, A. (2014). 'Regional resilience: an agency perspective'. *Regional Studies*, 48: 923–935.
Bürkner, H.J. (2010). 'Vulnerabilität und Resilienz: Forschungsstand und sozialwissenschaftliche Untersuchungsperspektiven'. *Leibniz Institute for Regional Development and Structural Planning (Erkner). Working Paper*, 43. Available at: https://leibniz-irs.de/fileadmin/user_upload/IRS_Working_Paper/wp_vr.pdf

Byrne, D. (1998). *Complexity theory and the social sciences: an introduction*. London: Routledge.
Chandler, D. (2014a). *Resilience: the governance of complexity*. Abingdon: Routledge.
Chandler, D. (2014b). 'Beyond neoliberalism: resilience, the new art of governing complexity'. *Resilience*, 2(1): 47–63.
Cohen, M.D., March, J.G. and Olsen, J.P. (1972). 'A garbage can model of organisational choice'. *Administrative Science Quarterly*, 17(1): 1–25.
Corning, P.A. (2002). 'The re-emergence of "Emergence": a venerable concept in search of a theory'. *Complexity*, 7(6): 18–30.
Davoudi, S. and Strange, I. (2009). 'Space and place in twentieth-century planning: an analytical framework and an historical review', in S. Davoudi and I. Strange (eds), *Conceptions of space and place in strategic spatial planning*. Abingdon: Routledge, pp. 7–42.
Davoudi, S., Shaw, K., Haider, L.J., Quinlan, A.E., Peterson, G.D., Wilkinson, C., Fünfgeld, H., McEvoy, D. and Porter, L. (2012). 'Resilience: a bridging concept or a dead end?'; '"Reframing" resilience: challenges for planning theory and practice interacting traps: resilience assessment of a pasture management system in northern Afghanistan'; 'Urban resilience: what does it mean in planning practice?'; 'Resilience as a useful concept for climate change adaptation?'; 'The politics of resilience for planning: a cautionary note'. *Planning Theory and Practice*, 13(2): 299–333.
French, S., Barchers, C. and Zhang, W. (2015). *Moving beyond operations: leveraging big data for urban planning decisions*. CUPUM Conference Paper 194. Cambridge.
Goldstein, B.E., Wessells, A.T., Lejano, R. and Butler, W. (2015). 'Narrating resilience: transforming urban systems through collaborative storytelling'. *Urban Studies*, 52(7): 1285–1303.
Gong, H. and Hassink, R. (2017). 'Regional resilience: the critique revisited', in T. Vorely and N. Williams (eds), *Creating resilient economies: entrepreneurship, growth and development in uncertain times*. Cheltenham: Edward Elgar, pp. 206–216.
Habermas, J. (1968). *Technik und Wissenschaft als "Ideologie"*. Frankfurt a.M.: Suhrkamp Verlag.
Healey, P. (ed.) (2010). *Making better places: the planning project in the twenty-first century*. Basingstoke: Palgrave.
Hill, E., Wial, H. and Wolman, H. (2008). 'Exploring regional economic resilience'. *Working Papers, Institute of Urban and Regional Development UC Berkeley*, 4. Available at: http://iurd.berkeley.edu/wp/2008-04.pdf
Hillier, J. (2010). 'Strategic navigation in an ocean of theoretical and practice complexity', in J. Hillier and P. Healey (eds), *The Ashgate research companion to planning theory: conceptual challenges for spatial planning*. Abingdon: Routledge, pp. 447–480.
Holling, C.S. (1973). 'Resilience and stability of ecological systems'. *Annual Review of Ecological Systems*, 4: 1–23.
Holling, C.S. (2001). 'Understanding the complexity of economic, ecological, and social systems'. *Ecosystems*, 4(5): 390–405.
Hudson, R. (2010). 'Resilient regions in an uncertain world: wishful thinking or a practical reality?' *Cambridge Journal of Regions, Economy and Society*, 3(1): 11–25.
Jessop, B. (1997). 'The governance of complexity and the complexity of governance: preliminary remarks on some problems and limits of economic guidance', in A. Amin and J. Hausner (eds), *Beyond markets and hierarchy: interactive governance and social complexity*. Cheltenham: Edward Elgar, pp. 111–147.
Jihad, F. and Jacques, T. (2012). 'Bricolage planning: understanding planning in a fragmented city', in S. Polyzos (ed.), *Urban development*. Rijeka, Croatia: Intech Publications, pp. 93–126.

Koch, M., Köhler, C., Othmer, J. and Weich, A. (2015). 'Planlos! Zur Einleitung', in M. Koch, C. Köhler, J. Othmer and A. Weich (eds), *Planlos! – Zu den Grenzen der Planbarkeit*. Paderborn: Wilhelm Fink, pp. 7–17.

Krätke, S. (2014). 'Cities in contemporary capitalism'. *International Journal of Urban and Regional Research*, 38(5): 1660–1677.

Kreibich, R. (1986). *Die Wissenschaftsgesellschaft: Von Galilei zur High-Tech-Revolution*. Frankfurt a.M.: Suhrkamp Verlag.

Lang, T. (2009). *Institutional perspectives of local development in Germany and England: a comparative study about regeneration in old industrial towns experiencing decline*. Potsdam: Universität Potsdam. Available at: http://opus.kobv.de/ubp/volltexte/2009/3734/

Luhmann, N. (1984). *Soziale Systeme: Grundriß einer allgemeinen Theorie*. Frankfurt a.M.: Suhrkamp Verlag.

Luhmann, N. (1997). *Die Gesellschaft der Gesellschaft*. Frankfurt a.M.: Suhrkamp Verlag.

MacKinnon, D. and Derickson, K.D. (2013). 'From resilience to resourcefulness: a critique of resilience policy and activism'. *Progress in Human Geography*, 37(2): 253–270.

Manson, S. (2001). 'Simplifying complexity: a review of complexity theory'. *Geoforum*, 32: 405–414.

Martin, R. and Sunley, P. (2015). 'On the notion of regional economic resilience: conceptualization and explanation'. *Journal of Economic Geography*, 1: 1–42.

Mykhnenko, V. (2016). 'Resilience: a right-wingers' ploy?', in S. Springer, K. Birch and J. MacLeavy (eds), *Handbook of neoliberalism*. Abingdon: Routledge, pp. 190–206.

Olsson, P., Gunderson, L.H., Carpenter, S., Ryan, P., Lebel, L., Folke, C. and Holling, C.S. (2006). 'Shooting the rapids: navigating transitions to adaptive governance of social-ecological systems'. *Ecology and Society*, 11(1): 18.

Portugali, J. (2011). *Complexity, cognition and the city*. Heidelberg: Springer.

Ratter, B. (2012). 'Complexity and emergence: key concepts in non-linear dynamic systems', in M. Glaser, G. Krause, B. Ratter and M. Welp (eds), *Human–nature interactions in the Anthropocene: potentials of social-ecological systems analysis*. New York: Routledge, pp. 83–101.

Reimer, M., Getimis, P. and Blotevogel, H. (2014). *Spatial planning systems and practices in Europe: a comparative perspective on continuity and changes*. New York: Routledge.

Rittel, H.W.J. and Webber, M.M. (1973). 'Dilemmas in a general theory of planning'. *Policy Sciences*, 4: 155–169.

de Roo, G. (2010). 'Planning and complexity: an introduction', in G. de Roo, and E.A. Silva (eds), *A planner's encounter with complexity*. Farnham: Ashgate, pp. 1–18.

de Roo, G. and Silva, E.A. (eds) (2010). *A planner's encounter with complexity*. Farnham: Ashgate.

Rydin, Y. (2010). 'Actor-network theory and planning theory: a response to Boelens'. *Planning Theory*, 9(3): 265–268.

Simmie, J. and Martin, R. (2010). 'The economic resilience of regions: towards an evolutionary approach'. *Cambridge Journal of Regions, Economy and Society*, 3(1): 27–43.

Taylor, N. (1998). *Urban planning theory since 1945*. London: SAGE.

Walby, S. (2007). 'Complexity theory, systems theory, and multiple intersecting social inequalities'. *Philosophy of the Social Sciences*, 37(4): 449–470.

Walker, B.H., Anderies, J.M., Kinzig, A.P. and Ryan, P. (2006). 'Exploring resilience in social-ecological systems through comparative studies and theory development: introduction to the special issue'. *Ecology and Society*, 11(1): 12.

Welsh, M. (2014). 'Resilience and responsibility: governing uncertainty in a complex world'. *Geographical Journal*, 180(1): 15–26.

10

RESILIENCE AND JUSTICE: PLANNING FOR NEW YORK CITY*

Susan S. Fainstein

Hurricane Sandy (often called Superstorm Sandy) swept over the New York region in October 2012. The Bloomberg administration summarised the storm's impacts on New York City as resulting in: '43 deaths … 6,500 patients evacuated from hospitals and nursing homes … Nearly 90,000 buildings in the inundation zone … 1.1 million New York City children unable to attend school for a week … close to 2 million people without power … 11 million travelers affected daily … $19 billion in damage' (New York City, Mayor's Office, 2013, page 11). Although the city's government had long been aware of its vulnerability to rising sea levels, Sandy dramatically highlighted the need for more attention to measures that would protect it and facilitate recovery from future threats. New York City is, after all, a land mass consisting of two islands and two peninsulas, making the threat to its inhabitants more extreme than in most coastal cities. Even before Sandy, Mayor Michael Bloomberg's administration had been gradually implementing plans to address storm danger: new construction conformed to more stringent standards for flood protection, while public investment had made improvements to basic infrastructure and to buffering of the water's edge. The impact of the storm, however, revealed their insufficiency, stimulated intensified planning and prompted identification of areas of the city particularly at risk.

The city government's increased concern with environmental vulnerability coincided with heightened attention among policy analysts to the consequences of climate change for densely settled areas. Instrumental in stimulating initiatives that would heed the warnings from scientists regarding sea level rise was a funding programme, launched by the Rockefeller Foundation in 2013, entitled *100 Resilient*

---

* This chapter was originally published as Susan S. Fainstein, "Resilience and Justice: Planning for New York City," *Urban Geography*, 2018 (https://doi.org/10.1080/02723638.2018.1448571).

Cities. As well as publicising the concept of resilience through publications and conferences, the foundation awarded grants to support 'chief resilience officers' in chosen cities, including New York. One major intent of the Rockefeller programme was to enlarge the meaning of resilience to include social and economic as well as environmental concerns. In the words of Judith Rodin (2014, page 7), the foundation's president, 'resilience building must move forward on three fronts: structural, social, and natural ... We need to strengthen and improve our approaches to governance and leadership, knowledge creation, communication, community development, and social cohesion.' The programme's web page asserts that resilience requires cities to be prepared for 'stresses [that] include high unemployment; an overtaxed or inefficient public transportation system; endemic violence; or chronic food and water shortages' (100 Resilient Cities, 2016).

This chapter explores New York City's recent policy initiatives developed under the rubric of resilience. In addition, it analyses whether the Rockefeller Foundation's definition of the term contributes to more just planning as it applies the concept to issues beyond preparedness for environmental disturbances. When resilience becomes a portmanteau for policies aimed at a more just, better governed and economically prosperous city, does it continue to have any real meaning? It also examines the implications of the now favoured strategy of 'making room for water'. Finally, it discusses the role played by community groups organised for environmental justice.

New York's Resiliency Planning

Mayor Michael Bloomberg established the Mayor's Office of Long Term Planning and Sustainability in 2006 and in the following year released *PlaNYC*, subtitled *A greener, greater New York* (New York City, Mayor's Office, 2007). Thereafter the city periodically released sustainability and resiliency reports and published an updated plan in 2011 with the title *A stronger, more resilient New York* (New York City, Mayor's Office, 2011).[1] The original *PlaNYC* emphasised sustainable growth in the context of increasing population and economic expansion. In 2013, responding to the depredations of Sandy, the mayor seized on the theme of resilience.[2] The amended blueprint in the 2013 version focused more on threat and made further recommendations including, inter alia, hardening the coastline through building dunes and nourishing beaches for creating open spaces for water absorption; and prescribing changes to the building code that would raise structures in areas prone to flooding, remove mechanical systems from basements and create a more durable power network. The document is notable for its specificity and detail. Although it does direct attention to low-income areas, it does not give them priority.

The 2013 plan declares that environmental planning should promote economic development. Thus, it proposes a levee along the lower east side tip of Manhattan adjacent to Wall Street that would extend the island into the East River and allow the development of a large mixed-use development on top of it (New York City, Mayor's Office, 2011, page 385). Although the mega-project idea disappears in later iterations, other Bloomberg economic initiatives already under construction occupy low-lying waterfront land. These include the development of a technology research centre on Roosevelt Island in the East River, located on a narrow site surrounded by water on three sides, and the massive Hudson Yards project stretching along the river on the west side of Manhattan. Rather than adopting the Rockefeller Foundation's broad definition of resilience, Bloomberg's new plan, like the previous versions, restricted its recommendations to the physical environment.

Waterside businesses and housing represent a challenge for resilience planning. Although the Bloomberg plan embodied certain principles of adaptation to risk in terms of creating channels and wetlands for water absorption, it did not, except for a small, state-sponsored programme on Staten Island, propose relocating occupants of the water's edge: 'In pledging to rebuild damaged or destroyed structures and infrastructure, the city disavowed reliance on another recognized approach to preparing for weather disasters: managed retreat from coastal areas that are particularly vulnerable to flooding and other storm-related damage' (McCardle, 2014). Yet, due to the placement of much public housing in waterfront areas at a time when port and industrial uses made waterfront residence undesirable, the city's substantial stock of public housing had been especially affected by Sandy. Consequently, its low-income residents were particularly vulnerable to future storms. Also seriously at risk were the city's waterfront industrial areas, where concentrations of hazardous facilities typically adjoined working-class communities, raising the potential for flooding to jeopardise public health as toxic materials mixed with the flood waters (see Bautista et al., 2015).

Mayor Bloomberg's successor, Bill de Blasio, who took office in 2015, inveighed against the previous administration's trickle-down approach to building wealth. In his election campaign De Blasio stressed that the veneer of New York's apparent prosperity disguised a tale of two cities, one wealthy represented by Bloomberg ('the one percent') and one struggling. He contended that his predecessor ignored the poor and middle class, instead focusing on Manhattan, the richest borough, and promoted programmes that primarily benefited the upper stratum. Bloomberg, in a much-quoted interview in *New York Magazine* (Smith, 2013), had asserted that: 'If we can find a bunch of billionaires around the world to move here, that would be a godsend, because that's where the revenue comes to take care of everybody else.' De Blasio in contrast called attention to the city's increasing income inequality and the shrinking of its middle class. In line with this way of thinking, he entitled his plan *One New York: the plan for a strong and just city* (*OneNYC*; New York City, Mayor's Office, 2015a).

The de Blasio plan enumerates four goals: economic growth, equity, sustainability and resiliency. Thus, it conforms to the Rockefeller Foundation's expansive conception of resilience. The Mayor's Office of Sustainability and the Mayor's Office of Recovery and Resiliency, with Rockefeller Foundation support, manage the development, implementation and progress of *OneNYC*. The plan builds on Bloomberg's template; its principal break with past approaches stems primarily from its broader scope and rhetorical emphasis on equity. The mayor presented it to the public in the South Bronx, New York's most impoverished area, surrounded by leaders of the city's environmental justice movement. In his press release he stated:

> Environmental and economic sustainability must go hand in hand – and OneNYC is the blueprint to ensure they do ... We are laying out specific goals to make sure that as we build a stronger, more sustainable, and more resilient city, we are also creating a more equitable one [including] ... our unprecedented goal of lifting 800,000 New Yorkers out of poverty.
>
> *(New York City, Mayor's Office, 2015b)*

The New York City Environmental Justice Alliance (NYC-EJA) praises the plan but also lists a series of actions that would give it more substance (NYC-EJA, 2016, page 3). Even while acknowledging the concerns of the environmental justice movement, which contests the placing of unwanted uses in poor communities, it contains few recommendations that would remove them. Despite its generalities about justice, specific proposals presented in the plan tend to favour the more affluent parts of the city. Moreover, like all New York mayors in the post-fiscal crisis era (i.e. since the mid-1970s; see Fainstein, 2001), de Blasio looks to the real estate industry as a driving force in producing economic development and increasing housing supply. Thus, he encourages further high-rise office production along with the upzoning of residential areas so as to provide incentives for building luxury housing that would cross-subsidise affordable housing construction. In his desire to increase housing supply he has encouraged waterfront development:

> Rather than promote a gradual retreat from the waterfront, the de Blasio administration has focused on making the city's coastlines, and its structures, more resilient to storms ... [The Sheepshead Bay neighborhood], like many others wedged against city coastlines, should see sea level rise by 11 to 21 inches by midcentury ... [According to expert on New York ecology] Ted Steinberg ... 'There is this short-term, profit-driven real estate logic in conflict here with an extremely likely long-term trend toward environmental instability.'
>
> *(Barkan, 2016; see also McPherson et al., 2014, page 511)*

Also following in the path of his immediate predecessor, de Blasio foresees 'creative' industries, especially those based on internet-related technologies, along with tourism, as essential to New York's future vitality. Although these industries have recently flourished in New York, adding significantly to the diversity of its economic base, their direct contribution to the wellbeing of low-income households is debatable. On the other hand, de Blasio has shown more interest than Bloomberg in preserving manufacturing, which clearly benefits poorer residents. Its preservation, however, exposes nearby neighbourhoods to environmental hazards, causing some communities to demand the removal of manufacturing facilities and raising difficult issues for reformers seeking justice in both labour markets and residential areas (Bautista et al., 2015, page 835).

As of this writing it is difficult to determine the extent to which the promises of *OneNYC* are being fulfilled. The city's success after Sandy in receiving $9.2 billion in federal money for recovery has meant that financial constraints have lessened compared to the past. Although one would hardly argue that New York was fortunate in sustaining three crises in the beginning years of the 21st century, they did precipitate a large inflow of federal funds. In contrast to New York's 1975 fiscal crisis, when the national government was notoriously begrudging in its response, its reaction to the 2001 attack on the World Trade Center, its floating of financial institutions after the 2007–8 financial crisis, and the Federal Emergency Management Agency's support after Sandy all substantially aided the city's finances.[3] This outside funding combined with a major expansion of the city's economy and tax revenues from very high-end real estate transactions meant that de Blasio had substantially more money to work with than any previous mayor.

The Mayor's Office issued a progress report in 2016 listing its achievements, but, while the report itemises many schemes planned or underway, the distribution of funds across the city is not clearly shown. Maps delineating coastal protection projects (New York City, Mayor's Office, 2016, pages 168–9), for which the city received over $600 million in federal grants in addition to the use of its own capital funds, indicates $453 million committed to strengthening the southern Manhattan coastline and smaller amounts allocated to a miscellany of projects in the other boroughs. Instead of the Bloomberg proposal for a mega-project on Manhattan's east side, the plan calls for a levee topped by parkland. All of the Bronx rates only $68 million, with the bulk ($45 million) going to buffer the Hunt's Point peninsula, where the region's food distribution centre is located.[4] Over $135 million is earmarked for Staten Island. The Queens and Brooklyn beaches and barrier islands along the Atlantic coast, which are especially vulnerable to sea level rise, are apportioned $168 million. Overall, the wealthy areas of Manhattan exceed the entire remainder of the city in terms of already committed financing for storm protection, although there are other projects listed in the study that are in the planning stages with no dollar sums yet specified.

In regard to the social and economic goals described in *OneNYC*, the report mentions many accomplishments but does not show them in a way by which the reader can ascertain their distributional impacts. Some of them, like offering universal pre-school and raising the minimum wage for city and contract employees, clearly direct benefits to the relatively disadvantaged. Others, like transit improvements, may increase equity, but since the figures provided are aggregated for the whole city, one cannot determine where the impact will be felt. While progress was made in providing affordable housing, the report does not show the extent to which these were net additions to the supply, since simultaneously other units were being withdrawn.[5]

Is the Term 'Resilience' Useful in Promoting Justice?

Under the Bloomberg plan the term resilience was largely restricted to policies directed at dealing with environmental threat.[6] Critics disparaged this approach as disregarding questions of justice and ignoring social issues. De Blasio, in contrast, extended the concept to encompass not just disaster preparedness but also 'strengthening community, social, and economic resiliency' (New York City, Mayor's Office, 2015a, page 222) by, inter alia, strengthening community-based organisations, supporting small businesses and local community corridors, and making workforce development part of all resiliency investments. *OneNYC*, in addition to its section on resiliency, also contains a potpourri of proposals ranging from improving early childhood education to increasing access to healthcare. In general, the plan is a compendium of good intentions; in the words of Mayor de Blasio,

> With this work, we will be prepared for the shocks and stresses ahead, and have the ability to bounce back stronger. OneNYC is ambitious, setting clear and aggressive goals. Our initiatives address every aspect of life in New York City – how we live, work, learn, and play, raise our children, and enjoy all our city has to offer.
> *(New York City, Mayor's Office, 2015a, page 3)*

The question arises as to whether using the term 'resilience' to cover so many laudable objectives disguises the trade-offs involved among them and the resulting distributions of costs and benefits. For example, retreat from the water's edge within inhabited areas results in displacement. Whether those displaced will receive sufficient recompense to acquire comparable dwellings, whether they will be able to replicate the community to which they formerly belonged and who will bear the financial costs of their move are controversial issues that produce winners and losers. Within the Rockefeller Foundation formulation, every challenge produces a win-win situation, but – as is indicated by neighbourhood opposition to siting industrial facilities nearby even when community members may benefit from employment within them – trade-offs are inevitable.

A highly publicised book by Bruce Katz and Jennifer Bradley (2013) of the Brookings Institution reflects this obliviousness to structural conflict and business power, revealing a Panglossian view of consensus building:

> It is clear that the real, durable reshaping [of America] is being led by networks of city and metropolitan leaders – mayors and other local elected officials, for sure, but also heads of companies, universities, medical campuses, metropolitan business associations, labor unions, civic organizations, environmental groups, cultural institutions, and philanthropies … [They are] using business planning techniques honed in the private sector. They are remaking their urban and suburban places as liveable, quality, affordable, sustainable communities and offering more residential, transport, and work options to firms and families alike.
> *(Katz and Bradley, 2013, page 3)*

From this perspective there are no structural conflicts within metropolitan areas and cooperation among all the various interests – capital and labour, white and black, industrialists and environmentalists – will ensure resilience, sustainability and economic development. Even while the de Blasio plan begins with a letter from the mayor identifying the threat of rising social inequality, it echoes this optimistic outlook by setting as a goal making 'every city neighborhood … safer by strengthening community, social, and economic resiliency' (New York City, Mayor's Office, 2015a, page 215). Yet, some groups inevitably will bear losses, choices cannot always be based on consensus and directing resources to the most vulnerable while placing undesired uses within affluent areas will provoke fierce resistance.

In one case, where the Bloomberg administration actually forced the Upper East Side of Manhattan to accept a waste-transfer station, the battle over its siting continued for more than 10 years and led to 11 court decisions before the city could begin construction (Hawkins, 2013). In response to the argument that Manhattan should bear its fair share of unwanted uses, the executive director of an adjacent sports complex sarcastically declared: 'You have a terrible problem in some communities, so let's put that problem somewhere else … That's supposed to represent borough equity?' (quoted in Gregory, 2014). Finally, in order to diminish the anguish of influential Upper East Side residents, the city agreed to build a new ramp to service the station and, unlike those in the other boroughs, to completely enclose it, thus adding hundreds of millions of dollars to its cost.[7]

Advocates of environmental justice have largely applauded the expansion of the term 'resilience' to virtually every policy area. The rationale for reformers using the term seems to be that, since resilience is a concept that offends no one, it can be used as a kind of Trojan horse to promote greater equity. The danger is that, in seeking to avoid confrontation through this stratagem, proponents of redistribution can lose their cause in the pressure to attain acceptability. An article by Portney and Berry (2016, page 201) inadvertently points to the likelihood of co-optation:

> The literature on urban economic development suggests a role for environmental lobbies that is compatible with, rather than antagonistic to, a growth orientation … Change in the urban business sector is … reflected in the push by many local Chambers of Commerce for more sustainability programs and policies … groups active in policymaking, including major businesses that once blocked any sort of environmental protection, have essentially been replaced by groups that are supportive of sustainability.

Although a pro-growth agenda does not necessarily cut against equity, it carries with it the potential for gentrification. Investment in facilities like bicycle paths, light-rail systems and village-like commercial centres that attract the creative class,[8] while not in themselves objectionable, absorb funds that could be directed toward bus transport and community centres in low-income areas. The argument frequently made that economic development is necessary to generate a surplus available for distribution to disadvantaged populations describes a possibility, but if business groups dominate policy making, then the likelihood of it being used for this purpose is limited.

Adaptation to Threat

A favoured approach in planning for resilience is to accommodate threat rather than prevent disturbances. This is in line with the ecological approach's stress on adaptation. The Dutch have pioneered the strategy of 'making room for water', which involves allowing flooding rather than using barriers to protect low-lying land (see Metz and van den Heuvel, 2012). Of course, there is really nothing new about this tactic except within the context of a country that previously relied on massive public works to fend off the surrounding seas. In fact, less developed countries have traditionally relied on annual flooding as the basis for agricultural productivity, and for decades wetlands protection has been a part of coastal zone management in many countries. Therefore, it is its social-historical positioning that makes the approach novel. We hear calls for dismantling existing 'hard' barriers to water in the United States, where the Army Corps of Engineers is taking down some dams and rebuilding wetlands in the Mississippi Delta. When we are speaking of unbuilt areas, little harm will be done, but even there some land owners will benefit while others will lose out. In cities the potential hardships are much greater when inhabited neighbourhoods are marked for inundation. Moreover, even the Dutch will continue to rely primarily on engineered barriers to water flows and the use of high technology-based emergency responses; they are modifying rather than leaving behind the mastery-of-nature model.

Since the most environmentally challenged urban land is typically inhabited by low-income residents who initially had few choices, returning the land to its pre-inhabited state places the cost burden of relocation on those least able to sustain it. Where waterfront land has been colonised by upper-income residents seeking views, the effort has largely been to protect them and keep them in place. Hypothetically, a poor community could be moved 'en bloc' to a more salubrious area, but this approach is very costly and seldom applied to marginalised communities. Simple compensation to individual households for the loss of their land would not supply the amount of money needed for former residents to settle in a decent home in more environmentally beneficial surroundings, nor would it reconstruct the community relations that had been severed. This situation, within the standard view of social ecology, is simply a dilemma of governance; within a more radical theory it is the consequence of capitalism under neoliberalism, where the resources to support everyone in a decent home and suitable living environment are withheld.

The strategy of making room for water is appealing and seems progressive, as it contravenes earlier, technologically based schemes. In areas not already heavily built up, it appears cost free. Yet even in these instances it can favour the more well-to-do over the poorer. An analysis of resilience planning in eight cities around the world finds that:

> Adaptation interventions can reinforce historic trends of socioeconomic vulnerability, compound patterns of environmental injustice, and create new sources of inequity. We find that unjust climate adaptation planning is not merely defined as a neglect of marginalized communities; rather, injustice should also be theorized and examined relationally (and spatially) against interventions in other – often more privileged – communities, sectors, and urban spaces. Rather than relying on technocratic or apolitical (or post-political) approaches often found in land use, infrastructure, or sustainability planning, climate adaptation plans must take into account historic legacies of social and racial injustice in order to avoid turning adaptation into a private and privileged environmental good with exclusionary and maladaptive externalities.
>
> *(Anguelovski et al., 2016, page 13)*

Anguelovski et al. (2016) show how sins of both commission (e.g. protecting only economically valuable areas) and omission (e.g. failure to retreat from vulnerable areas) characterise land-use planning for climate adaptation. A study of three ocean-side counties in Florida further illustrates the point (Puszkin-Chevlin, 2007). It shows that the wealthiest county, characterised by 4 acre lots and ample open land, could much more easily accommodate flooding than its needier neighbours. In the latter, which were desperate for tax revenues, and where the most desirable properties were at the water's edge, economic development programmes called for high-density construction in flood-prone areas.

The technocratic approach to planning for resilience generally is based on modelling and conducted as an exercise in risk assessment followed by a cost-benefit analysis of alternative responses to determine 'acceptable risk' (Fischer, 1991; see also Welsh, 2014). Risk calculations, however, cannot tell us what level of risk is tolerable nor do they break down the question into that of risk for whom. Instead they aim at giving precise numbers despite the actual uncertainty involved:

> The clearest message from the changing evidence base over the last decade concerns the dangers of false precision ... With regard to flooding, the data appears to be particularly subject to rapid and fundamental change and raises questions as to the extent to which it can be distilled to a probabilistic figure or clear spatial delineation between 'safe' areas and those 'at risk'.
>
> *(White, 2013, page 110)*

These numbers, however, are demanded by planners so that they can decide on desirable levels of density, and insurers so that they can develop underwriting criteria and calculate premiums (see Grove and Pugh, this volume, for a discussion of how insurance bolsters neoliberal order and transfers local risk to profit-making opportunities in global financial markets). They fit into the current fad for 'evidence-based planning'. The discourse of evolutionary resilience that underlies support for adaptation instead of prevention, the apparent scientific precision of risk analysis and the glamour of complexity theory allow conversations that do not confront the real issue of which groups of the population will benefit from the expenditure of public resources. These conversations avoid divisiveness by assuming that everyone will gain if resilience is enhanced, and the allusion to the great complexity involved in achieving resilience creates a cloud of obfuscation around the question of who is getting what.

Cote and Nightingale (2012), in an insightful article, complain that studies of vulnerability mask the necessity to ask normative questions. Nevertheless, in an otherwise critical discussion, they praise the interdisciplinary nature of resilience analyses that develop the concept of a social-ecological system (page 477). They comment that 'one promising aspect of this work is the genuine commitment to a holistic approach that integrates a diversity of scholarly disciplines and embraces complexity'. Yet, although one cannot dispute the merits of an interdisciplinary approach, and certainly bringing together physical and social sciences is rare and desirable, the emphasis on complexity feeds into the very aspects of resilience planning that Cote and Nightingale dislike. Rather than invoking 'Occam's razor', proponents of complexity theory, enchanted by the computer's capacity to describe thousands of interactions, fail to understand the importance of social power (see Flyvbjerg, 1998).

Cote and Nightingale are among a number of theorists who, when discussing the paradigm of social ecology from which the argument for resilience derives (see Gunderson and Holling, 2002), critique it for inadequately addressing questions of political power and the role of the state and for incorporating a conservative political bias (see e.g. Swanstrom, 2008; Wilkinson, 2012). According to this general argument, the social ecology perspective regards the interaction of humans and nature as comprising a multitude of variables with no identifiable dominant causal agent. Rather, everything is related to everything else. Scholarly work derived from this viewpoint turns to complexity theory and complicated modelling to determine levels of risk. Yet, examining social phenomena through the lens of complexity theory leaves the analyst with enormous mapping jobs and model-building challenges but provides little in the way of decision rules. Thus, Eric Swyndegouw (2010, page 303) comments: 'Unforeseen changes are seen either as the effect of "externalities" … or as a catastrophic turbulence resulting from initial relations that spiral out in infinitely complex and greatly varying configurations such as those theorised by Chaos or Complexity Theory'. For Swyndegouw this perspective amounts to denying the relationships of domination that characterise the hidden, conservative ideology within environmentalism.

A recent attempt to deal with climate change in the New York region illustrates the power of capitalist interests in discussions of environmental measures (Stile, 2016). The state of New Jersey has the second lowest rate of taxation on gasoline in the United States. A bill in the New Jersey legislature to raise the tax was blocked by legislators who opposed any form of tax increase. A compromise plan supported by the governor proposed to reduce the sales tax as compensation, while Republicans in the legislature suggested the elimination of the estate tax, which would affect only the wealthiest residents. As of this writing no agreement was reached, but the various approaches indicate that environmental policy had to conform to a neoliberal, anti-tax regime for any kind of success.

Neoliberalism, which promotes low taxes, minimal regulation, privatisation and market-based solutions, has been foundational to policy making for resilience. In attempts to evade regulatory restrictions or raising taxes on contributors to global warming, 'cap and trade' systems, which establish markets in rights to pollute, have become the preferred method for reducing greenhouse gases and acid rain. Proposals that require spending a great deal of money on poor people are generally regarded as inefficient as well as politically impossible and therefore are evaded. Based in neoclassical economic thought, neoliberalism's starting premises are that society consists of individual consumers and producing/investing firms. Consequently:

> We arrive at a view by which the best way to govern society is through a greater awareness of our own [individual] behaviour. Indeed, a major claim here is that the way resilience works, certainly in Anglo-Saxon approaches, is to move fairly swiftly from thinking about the dynamics of systems to emphasising individual responsibility, adaptability and preparedness.
>
> (Joseph 2013, page 40)

Given the limited resources available to city governments, it is inevitable that they look to the private sector for financing resilience efforts, but this requires that the programmes developed be acceptable to funders.

Much of the literature on urban politics has identified a growth machine that powers property development (Logan and Molotch, 1987) and urban regimes skewed toward business interests. The move to create resilience converges with these interests in developing parks that will absorb flooding from nearby waterways. The value of these parks for real estate developers is that they provide views and amenities for nearby residents. Within New York recently constructed parks along the Hudson (the High Line, Hudson River Park and Battery Park City) and East Rivers (Brooklyn Bridge Park) operate under the aegis of conservancies governed by boards that operate independently of city government. They have stimulated the investment of many billions of dollars in adjacent areas and corresponding profits to developers. Waterside parks in poor neighbourhoods, on the other hand, have little appeal to private interests. To some extent the de Blasio administration has attempted to rectify this by directly investing in a number of neglected parks in unfashionable parts of the city and calling for the conservancies to contribute to their maintenance. In his Community Parks Initiative (CPI) de Blasio identified 35 parks in impoverished neighbourhoods to receive priority in capital spending for improvements. Nevertheless, according to the NYC-EJA (2016, page 26): 'while on the surface CPI seems like a benign attempt to improve parks in low-income neighborhoods, it may have unanticipated adverse effects'. Because of the selection criteria used in determining the CPI zones, the Environmental Justice Alliance worries that the criterion of population growth for the chosen neighbourhoods could 'reward rapidly gentrifying communities with increased park investments, thereby creating a wicked loop where gentrifying communities attract a disproportionate share of park investments, which in turn spurs even more displacement' (NYC-EJA, 2016, page 26).

Countervailing Pressures

Community-based organisations concerned with environmental management have long pressed for greater democratisation of the decision-making process. Planning documents published by the mayor's office always insisted that communities were involved with plan formulation. Thus, Mayor Bloomberg declared, when initiating the post-Sandy process to create a more resilient New York: 'To succeed, the plans must include the input of the people who live and work in these communities – and they will. Members of the community will assist in shaping and implementing each community plan – and that will be just the beginning of our work' (quoted in New York City, Mayor's Office, 2013, page 5). Similarly, Mayor de Blasio asserted:

> *OneNYC* is based on ideas coming from thousands of New Yorkers. We asked civic, community, and business leaders what they thought we should be doing. We heard from everyday New Yorkers – at town hall meetings and online, in polls and surveys – who told us about what works and what could be better in their lives, and what they imagine for New York's future.
> *(New York City, Mayor's Office, 2015a, page 7)*

The extent to which the decision-making process has actually adhered to these norms, however, is questionable. The NYC-EJA praises the de Blasio initiative for requiring community-based brownfields planning but comments that 'the implementation of the City's sustainability and resiliency initiatives remains primarily top-down' (NYC-EJA, 2016, page 17). In particular, it contended that the plan did not respond to community-defined priorities. For example, Bronx community leaders have criticised the plan for giving priority to energy resilience when they regarded flood protection as their most significant need (Mock, 2016). More broadly citizens had no say in the overall allocation of resources by geographic area.

Within individual neighbourhoods local groups possess knowledge unavailable to centrally located officials. Jason Corburn terms local knowledge 'street science'. He argues that 'the kind of information that street scientists offer varies – from missing hazard information to detailed cultural practices that influence human exposures to pollution – but this knowledge is as much "expertise" as the information that professionals offer' (Corburn, 2005, page 201). While engineering aspects of dealing with climate-caused threats exceed the layman's capabilities, assessments of investment priorities within affected communities do not. Moreover, providing resources to community groups can allow them to hire consultants to advise on technical issues.

New York has a variety of advocacy groups that press for socially sensitive environmental policy. Many of them are based in individual neighbourhoods and perform bridging functions between their areas and the city, acting as both critics and sponsors of efforts at environmental improvement (Connolly et al., 2013). In addition, there are many city-wide organisations, ranging from elite establishment groups like the Municipal Arts Society to 'Green Guerrillas', which sponsor neighbourhood gardens throughout the city. The NYC-EJA is prominent as a coalition among advocacy organisations pushing for environmental quality. Founded in 1991, it is a non-profit, tax-exempt, city-wide network; members include low-income neighbourhood organisations and labour unions, and it links up with similar groups throughout the region. It was one of the conveners of the New York Regional Assembly, which in 2013 brought together representatives from 40 organisations in New York and New Jersey to develop strategies for the Sandy recovery process, and it has provided extensive critiques of the Bloomberg and de Blasio resilience plans.

NYC-EJA receives foundation funding, including from the Rockefeller Foundation, and has been able to collaborate with various university research teams in order to develop technical analyses. In addition, it has on staff several individuals with technical expertise. It is led by Eddie Bautista, a long-time community advocate, who previously served as director of Mayor Bloomberg's Office of Legislative Affairs. The organisation focuses on environmental justice and attempts to influence government at the city, state and national levels. As is obvious from its ties to foundations and the city government, it is not regarded simply as an antagonist; nevertheless, it has consistently defined its role as being an advocate for poor communities and communities of colour. It has organised marches, sent representatives to speak in hearings and built alliances. It works closely with The Point Community Development Corporation (CDC), a South Bronx CDC that focuses on economic development and culture as well as environmental justice. The Point is unusual among CDCs in that it is not a housing developer. One notable aspect of both NYC-EJA and The Point is that their concern extends beyond the residential community to employment and workplace safety. NYC-EJA has especially focused on the vulnerabilities of waterfront industrial areas, where the city had traditionally herded polluting industries and egregiously ignored infrastructure needs. As discussed earlier, low-income residents affected by nearby industry, even if the only impact arises from truck traffic, frequently oppose having manufacturing, distribution or waste disposal facilities nearby. These, however, are principal sources of employment for low-skilled workers. Mediating this conflict through devising solutions like non-polluting vehicles or buffer areas represents the kind of application of local knowledge that contributes to environmental justice. As noted by Grove and Pugh (this volume), such research can destabilise the political-economic inequality that promotes environmental injustice.

Conclusion

Within the United States environmentalism developed from the conservation movement of the late 19th and early 20th centuries. Based in upper- and especially professional-class viewpoints, it initially focused on building reservoirs to contain flood waters and attain efficient land management (Hays, 1959). Then as now technocratic thought dominated policy formulation: 'They [the conservationists] required new administrative methods, utilizing to the fullest extent the latest scientific knowledge and expert, disinterested personnel' (Hays, 1959, page 266). Justified in terms of the public interest, conservation was regarded as benefiting everyone, and its distributional consequences were ignored. At the same time, various wealthy families pressed for the preservation of scenic areas and were instrumental in the founding of the national park system under the administration of President Theodore Roosevelt (Foresta, 1984).

Whereas the conservation movement originally directed its attention to uninhabited areas, environmentalism more broadly has had a significant urban component, previously called sustainability and now frequently labelled resilience. The use in this book of the term resilience machine captures how its deployment can serve the interests of capital. Within New York City, mayoral administrations have reacted to the threat posed by rising sea levels through resilience planning aimed at protecting real estate investors and encouraging further property development. In other words, resilience planning, even while it has enhanced the city's ability to survive flooding, has contributed to the interests of the growth machine. It has, however, also provoked organisations dedicated to environmental justice to broaden their scope and affect the distribution of benefits. These processes and counter-processes can be detected elsewhere, as Grove and Pugh (this volume) illustrate in their discussion of the Caribbean, or Nelson et al. (2007) show in their depiction of New Orleans after Hurricane Katrina.

Whether intentional or not, environmental policy has had exclusionary effects and generally has served capitalist needs despite its progressive veneer (Harvey, 1996). The original national parks were not accessible to impoverished urban residents. The siting of undesirable facilities has been governed by the strength of political resistance and the price of land, resulting in the placement of toxic uses within low-income communities. Suburban home owners typically oppose the placement of affordable housing in their midst, rationalising their resistance in terms of sewer overburden, wetlands preservation and containment of run-off. Support for urban parks emanates from real estate interests whose properties gain value from nearby open space. By and large justice has not been central to the demands of environmentalists.

The significance of the environmental justice movement, embodied in organisations like NYC-EJA, has been to force recognition of the impact of conservation measures on low-income communities. The strength of NYC-EJA is its ability to provide resources and expertise, attributable to the existence of a paid staff rather than reliance only on volunteers (see Fainstein and Fainstein 1974, page 200 and following pages). This staff can bring people into the streets when deemed appropriate; it can also subject public decisions to technical analyses and maintain an ongoing scrutiny of governmental actions even when popular mobilisation is lacking.

The popularity of the term 'resilience' in the present epoch has not done away with the tension between expert-driven decision making, based on calculations of aggregate costs and benefits, and demands that aggregates be broken down into analyses of who gains and who loses. Some participants in this latter effort have called for resilience to encompass consideration of all aspects of urban life. The success, however, of the environmental justice movement has derived from its focus on a relatively limited spectrum of issues and its ability to develop expertise within this area. If the definition of resilience includes the ability to recover from all malaises including industrial obsolescence and physical disability, the term loses meaning. If, however, it is tightly connected to the interaction of disadvantaged groups with the physical environment, it has the capacity to contribute to greater justice.

Notes

1 The 2007 plan used the term sustainability rather than resilience. For critiques of it, see Cowett (2008); Marcuse (n.d.).
2 Although New York's zoning and site plan regulations are codified in enacted legislation, planning documents emanating from the Mayor's Office, like this one and subsequent iterations, constitute administrative guidance and do not represent binding commitments.
3 The New York City Public Housing Authority received almost $2.6 billion from the Federal Emergency Management Agency in capital funding, most of it as a one-time capital grant to aid recovery after Sandy (New York City Council, 2016).
4 The market, which serves the entire metropolitan area, is a major employer of unskilled workers (Fainstein and Gray, 1996). It abuts a working-class neighbourhood, which suffers from the negative air pollution effects caused by the trucks serving the market. Resulting tensions between economic and environmental goals recently flared when Fresh Direct, the city's largest grocery delivery service, desired to place its main distribution facility in the market.
5 In July 2016 the de Blasio administration announced that it was exceeding its affordable housing goals. Most of the housing, however, was simply prevented from leaving the rent-regulated stock rather than representing net additions (Bagli, 2016).

6 Parts of this section are drawn from Fainstein (2015).
7 According to the city's Independent Budget Office (IBO), the cost of continuing under the preceding arrangement whereby trucks removed Manhattan's waste to landfills outside the city was estimated at a present value of $233 million versus $633 million for use of the marine transfer station. The main reason for the larger cost was the capital expenditure involved from constructing the new facility. The IBO estimate was made before the agreement concerning the additional access ramp and thus did not include its $31 million estimated cost (New York City, Independent Budget Office, 2014). The waste would be moved by truck to the marine transfer station, where it would be transferred to barges. Thus, the reduction in air pollution caused by water transport would be obtained outside the city (Weaver, 2014).
8 Portney and Berry (2016, page 201) mention these kinds of amenities as investments that environmentalists and business groups both like.

References

100 Resilient Cities (2016). Available at: www.100resilientcities.org/about-us#/-_/
Anguelovski, I., Shi, L., Chu, E., Gallagher, E., Goh, K., Lamb, L., Reeve, K. and Teicher, H. (2016). 'Equity impacts of urban land use planning for climate adaptation: critical perspectives from the Global North and South'. *Journal of Planning Education and Research*, DOI: 0739456X16645166.
Bagli, C.V. (2016). 'De Blasio administration says it is ahead of schedule on affordable housing'. *New York Times*, 26 July.
Barkan, R. (2016). 'Transformation on Brooklyn's southern shore'. *New York Times*, 16 September.
Bautista, E., Osorio, J.C. and Dwyer, N. (2015). 'Building climate justice and reducing industrial waterfront vulnerability'. *Social Research*, 82(3): 821–838.
Connolly, J.J., Svendsen, E.S., Fisher, D.R. and Campbell, L.K. (2013). 'Organizing urban ecosystem services through environmental stewardship governance in New York City'. *Landscape and Urban Planning*, 109(1): 76–84.
Corburn, J. (2005). *Street science*. Cambridge, MA: MIT Press.
Cote, M. and Nightingale, A.J. (2012). 'Resilience thinking meets social theory: situating social change in socio-ecological systems (SES) research'. *Progress in Human Geography*, 36(4): 475–489.
Cowett, P. (2008). 'New York's sustainability plan: trailblazer or copycat?' Hunter College Center for Community Planning and Development, Working Paper no. 2, June.
Fainstein, S.S. (2001). *The city builders* (revised edn). Lawrence, KS: University Press of Kansas.
Fainstein, S.S. (2015). 'Resilience and justice'. *International Journal of Urban and Regional Research*, 39(1): 157–167.
Fainstein, S.S. and Fainstein, N.I. (1974). *Urban political movements*. Englewood Cliffs, NJ: Prentice-Hall.
Fainstein, S.S. and Gray, M. (1996). 'Economic development strategies for the inner city: the need for governmental intervention'. *Review of Black Political Economy*, 24(2–3): 29–38.
Fischer, F. (1991). 'Risk assessment and environmental crisis: toward an integration of science and participation'. *Organization and Environment*, 5(2): 113–132.
Foresta, R.A. (1984). *America's national parks and their keepers*. Washington, DC: Resources for the Future, distributed by Johns Hopkins University Press.
Flyvbjerg, B. (1998). *Rationality and power*, trans. S. Sampson. Chicago, IL: University of Chicago Press.

Gregory, K. (2014). 'Fight awaits de Blasio on opening Upper East Side trash transfer site'. *New York Times*, 4 April.
Gunderson, L.H. and Holling, C.S. (eds) (2002). *Panarchy: understanding transformations in systems of humans and nature*. Washington, DC: Island Press.
Harvey, D. (1996). *Justice, nature and the geography of difference*. Oxford: Blackwell.
Hawkins, A.J. (2013). 'Upper East Side won't hear defeat on waste-transfer station'. *Crain's New York Business*, 21 April.
Hays, S.P. (1959). *Conservation and the gospel of efficiency*. New York: Atheneum.
Joseph, J. (2013). 'Resilience as embedded neoliberalism: a governmentality approach'. *Resilience*, 1(1): 38–52. DOI: 10.1080/21693293.2013.765741
Katz, B. and Bradley, J. (2013). *The metropolitan revolution*. Washington, DC: Brookings Institution.
Logan, J. and Molotch, H. (1987). *Urban fortunes*. Berkeley, CA: University of California Press.
Marcuse, P. (n.d.). 'PlaNYC is not a 'plan' and it is not for "NYC"'. Hunter College Center for Community Planning and Development, Sustainability Watch Working Paper.
McCardle, A. (2014). 'Storm surges, disaster planning, and vulnerable populations at the urban periphery: imaging a resilient New York after Superstorm Sandy'. *50 Idaho Law Review*, 19.
McPhearson, T., Hamstead, Z.A. and Kremer, P. (2014). 'Urban ecosystem services for resilience planning and management in New York City'. *Ambio*, 43(4): 502–515.
Metz, T. and van den Heuvel, M. (2012). *Sweet and salt: water and the Dutch*. Rotterdam: NAi.
Mock, B. (2016). *Does the OneNYC sustainability plan really address equity?* Available at: www.citylab.com/politics/2016/04/does-the-onenyc-sustainability-plan-really-addres s-equity/477852/
Nelson, M., Ehrenfeucht, R. and Laska, S. (2007). 'Planning, plans, and people: professional expertise, local knowledge, and governmental action in post-hurricane Katrina New Orleans'. *Cityscape*, 9(3): 23–52.
New York City, Independent Budget Office (2014). 'Letter from Ronnie Lowenstein, Director, to Council Member Benjamin J. Kallos'. 21 October. Available at: www.ibo. nyc.ny.us/iboreports/2014e91stwtsLetter.pdf
New York City, Mayor's Office (2007). *PlaNYC: a greater, greener New York*. Available at: www.nyc.gov/html/planyc/downloads/pdf/publications/full_report_2007.pdf
New York City, Mayor's Office (2011). *PlaNYC: update*. Available at: www.nyc.gov/htm l/planyc/downloads/pdf/publications/planyc_2011_planyc_full_report.pdf
New York City, Mayor's Office (2013). *A stronger, more resilient New York*. Available at: http://s-media.nyc.gov/agencies/sirr/SIRR_singles_Lo_res.pdf
New York City, Mayor's Office (2015a). *One New York: the plan for a strong and just city*. Available at: www.nyc.gov/html/onenyc/downloads/pdf/publications/OneNYC.pdf
New York City, Mayor's Office (2015b). 'Press release: Mayor de Blasio releases one New York: the plan for a strong and just city'. 22 April.
New York City, Mayor's Office (2016). 'One NYC Progress Report'. Available at: www1.nyc.gov/html/onenyc/downloads/pdf/publications/OneNYC-2016-Pro gress-Report.pdf
New York City Council (2016). 'Report on the fiscal 2017 preliminary budget and the fiscal 2016 preliminary mayor's management report New York City'. Housing Authority, 28 March. Available at: http://council.nyc.gov/html/budget/2017/pre/NYCHA.pdf
New York City Environmental Justice Alliance (2016). *NYC climate justice agenda: strengthening the mayor's One NYC plan*. New York: NYC-EJA, April.

Portney, K.E. and Berry, J.M. (2016). 'The impact of local environmental advocacy groups on city sustainability policies and programs'. *Policy Studies Journal*, 44(2): 196–214.

Puszkin-Chevlin, A.C. (2007). *Determinants of local hazard mitigation policy and built environment vulnerability: three case studies from Florida's Treasure Coast.* PhD dissertation, Columbia University.

Rodin, J. (2014). *The resilience dividend.* New York: Public Affairs.

Smith, C. (2013). 'In conversation: Michael Bloomberg'. *New York Magazine*, 7 September. Available at: http://nymag.com/news/politics/bloomberg/in-conversation-2013-9/

Stile, C. (2016). 'N.J. gas-tax fight gets risky for legislators'. *Record*, 9 July.

Swanstrom, T. (2008). *Regional resilience: a critical examination of the ecological framework.* Working paper, Institute of Urban and Regional Development, University of California at Berkeley, 25 April.

Swyndegouw, E. (2010). 'Trouble with nature: "ecology as the new opium for the masses"', in P. Healey and J. Hillier (eds), *The Ashgate research companion to planning theory*. Farnham: Ashgate, pp. 299–318.

Weaver, S. (2014). 'City adds second ramp to marine transfer station plan, adding $30m to cost'. dna info, 6 August. Available at: www.dnainfo.com/new-york/20150806/yorkville/city-adds-second-ramp-marine-transfer-station-plan-adding-30m-cost

Welsh, M. (2014). 'Resilience and responsibility: governing uncertainty in a complex world'. *Geographical Journal*, 180(1): 15–26.

White, I. (2013). 'The more we know, the more we know we don't know: reflections on a decade of planning, flood risk management and false precision'. *Planning Theory and Practice*, 13(1): 106–113.

Wilkinson, C. (2012). 'Urban resilience: what does it mean in planning practice?' *Planning Theory and Practice*, 13(2): 319–324.

11

SEEKING THE GOOD (ENOUGH) CITY

Brendan Gleeson

Introduction

Much contemporary talk is of the resilient city. As this volume points out, the resilience agenda has asserted itself in many city discourses and outlooks with impressive, not to say machinic, force. Indeed, as Davoudi (2016) has pointed out recently, resilience in many settings presents as an alternative governance rationality in an age of risk and uncertainty (Beck, 2009) that contrasts with long embedded planning rationalities, especially those committed to the technocratic management of 'knowable' urban imperatives. At the 'governance coal face' new resilience frameworks are emerging alongside mainstream metropolitan planning systems, an uneasy and as yet largely unexamined juxtaposition of the old and new. For example, the new Resilient Melbourne (2016) strategy asserts institutional priority alongside the city's evolving and officially embedded *Plan Melbourne* (State of Victoria, 2017) metropolitan strategy. Both assert the necessity of good planning, albeit from different starting points, but what is the good of planning?

Considering its recent assertion and rising importance, what good is the resilient city if it isn't good? Good must surely mean a state where humans can flourish as natural social beings, not merely endure or survive. The current epochal crisis suggests species catastrophe to many – that is, no prospect for life, let alone goodness. This is not a progressive, or even realistic, view. A critical social scientific perspective rooted in Arendtian ideals, notably faith in the human facility for 'natality' (Arendt, 1998 [1958]), rejects catastrophism without neglecting the scale of epochal threat. Goodness as flourishing, not morality, is a prospect we can hold to. The modern urban project can be renewed, inevitably through the 'fires' of global change. In a time when critical science is at a low ebb (Sayer, 2011), where are the resources for renewal, especially for new urban imaginaries that break with the current constructions (smart city, green city, knowledge city, etc.) that 'normalise the apocalypse' (Žižek, 2010)?

Surprisingly, the secular and religious notions of the good city might not be so far apart. The biblical construction of humanity as pilgrims seeking the 'good enough' city has resonance for the secular tradition, especially now. In a time when our species has to 'hit the road' again, the idea of urban pilgrimage might bear reconsideration. This idea signposts resilience as a journey of re-creation, not a securitised end state. It enjoins us to seek the good enough city and not to wall ourselves in resilience. If a 'species journey' correctly describes what we must undertake in the face of awful environmental and social threats, then Davoudi's (2012) caution that some constructions of resilience may present dead ends not evolutionary pathways is important to bear in mind.

This chapter considers the prospects for renewal of urban imaginaries, especially the ambiguous ideal of resilience, which might emerge from encounters between religious and secular thought. It reviews a recent contribution by the theologian Luke Bretherton (2015) which essays the pathway to a 'good enough city' in perilous times. One idea that arises from this encounter is that of 'resilience without optimism', to borrow Eagleton's (2015) recent trope. This means to accept the testimony of reason, that capitalism faces dissolution, but also the testimony of faith embodied in Arendt's notion of natality. That is to say, human history is a pilgrimage never completed and subject to privations and catastrophes, but safeguarded by the certainty of rebirth. What cannot be discarded on the journey are the hard-won ideals of justice and democracy, which, as Fainstein (2015) points out, are too often masked in resilience thinking. This necessity enjoins us to not take up 'regressive resiliences' that see these values as excess historical baggage that weigh upon a narrower, more defensive view of the human prospect.

The Urban Age and Its Discontents

An urban age has been declared by global institutions and in expert commentary. A rapidly increasing majority of humanity now resides in urban settings, especially the fast-growing 'meso-cities' of the developing world. The major transnational institutions bestow great significance to urbanisation as a force shaping human fortunes (OECD, 2010; World Bank, 2010; United Nations Habitat, 2012). For the past half-decade, the United Nations (UN) has broadcast the message of a new urban ascendancy. UN Habitat (2012, page v) enthuses: 'A fresh future is taking shape, with urban areas around the world becoming not just the dominant form of habitat for humankind, but also the engine-rooms of human development as a whole'.

Triumphalism abounds. The banners of 'revolution' and 'triumph' have been unfurled by the leading urban enthusiasts. These offerings include *The Triumph of the City* by Harvard economist Ed Glaeser (2011) and *Welcome to the Urban Revolution* by the Canadian urban 'practitioner and thinker' Jeb Brugmann (2009).

Aerotropolis is ordained as 'the way we'll live next' by Kasarda and Lindsay (2011). There is no room for urban apostasy – *Cities are Good for You*, claims Hollis (2013), inviting that we applaud 'the genius of the metropolis'.

To be sure, the urbanisation project that has been central to modernisation has reached new heights of species' significance. It cannot be denied that urbanisation has been the 'long march' of human improvement (with all of the pitfalls suggested by the metaphor). And yet, despite the happiness of expertise, the new urban preponderance also marks a dangerous unravelling of human prospect. The testimonies of manifest environmental, social and economic breakdowns struggle to be heard above the chorusing of the urban age. 'The horsemen of the apocalypse' (Žižek, 2010) are the unheeded town criers of an endangered modernity. For Davis (2010), climate change is the great reckoning of the urban age. He deploys biblical metaphor – that of the Ark – to describe the human project of resilience in the face of imminent climate threat. This resonates with a progressive view of resilience as a journey whose compass settings must remain the values of justice and democracy not mere survival and enclosure (Fainstein, 2015).

It seems that influential parts of urban commentary are declaring an end of Promethean modernity. This term recalls Prometheus, the titan from Greek mythology who scorned the natural order and imagined himself greater than the gods. By express commitment, industrial modernity has been godless and boundless, without limits (Beck, 2009). The idea of self-limitation has been regarded as heresy; a primitive notion that conspires against the modern ideal. We might ask the question in the vexed debate on the 'Anthropocene', has the Promethean order finally run its course? Has nature put chains on us again? This time around our necks? This is to reprise an old question, raised for example by Illich (1977) who urged embrace of self-limitation as a guiding social principle as post-war capitalism encountered and suffered the inevitable contradictions of compound growth and technocratic assertion.

A danger to human prospect is the increased decoupling of urbanisation from its modern work of species liberation, captured in the old German refrain 'Stadtluft macht frei'. More and more it seems urbanisation is connected to the imperatives of accumulation, not humanisation. The compact city ideal of progressive urbanism has been redeployed to the work of (vertical) accumulation (Gleeson, 2014). In Western cities we see an urbanism that is hyper-intensive, hyper-accumulative but strangely impoverished. In a (long) word, 'hypertrophic urbanism'. Cities are being driven to higher density by the dictates of power and capital not sustainability. An urban world once again remade with shattering force by the barons of power.

In so-called developed cities austerity governance and a relentlessly growing security state increase their hold. We are in fact, from the point of view of democracy, witnessing a reversal of the development cycle in Western states. The politics of representative democracies, including urban politics, is largely quiet and impotent. They have no model to offer an anguished world. Will models of human survival come from non-Western countries and cities?

Much of the rest of the world is consigned to what Davis (2006) has called a *Planet of Slums*, an urbanism marred by appalling environmental pollution and social disparity. In South American cities, intensifying urban capital is hideously mocked by the rise of vertical favela. Harvey (2010) eloquently describes the predatory, frequently brutal, relations that govern much of contemporary slum life. Žižek (2010) believes that the slums of the South are restive with radical, even violent, revolutionary potential. This contrasts with the 'revolution' (meaning 'innovation unleashed') forecast by urban enthusiasts (e.g. Brugmann, 2009; Glaeser, 2011). Davis (2010, page 30) fears the great inundation of cities in the Global South that must follow a warming climate, warning that 'cities of poverty will almost certainly become the coffins of hope'. His counsel is not complete despair but that 'we must start thinking like Noah' (Davis, 2010, page 30). By contrast, the urban enthusiast Kahn (2010, page 241) in *Climatopolis* adopts a remarkably sanguine view of the warming crisis: 'our cities will thrive in the hotter future'. (To be sure, Kahn hedges his sanguinity, admitting, 'much of my optimism goes out the window if climate change inflicts abrupt shocks'.)

The bright mood of popular urban commentary contrasts with the terrible outlook facing much of the human species; now reconstituting itself as 'homo urbanis'. As Beck (2012, page 161) observed: 'It's no wonder that the business in hopelessness is flourishing'. And yet is this cause for despair – do we live in the ever lengthening shadow of inevitable catastrophe? Further, how to reconcile recognition of the gravity of the contemporary condition with the progressive commitment to human improvement and striving? Catastrophe may be the terrible spectre that invokes resilience response, but surely a problematical notion – and certainly no ideal! – for critical social science that refuses a naturalised, ordained human future. Catastrophism is surely as antithetical to human prospect as the urban triumphalism that happily trumpets a new age of human emancipation.

The Ruse of Catastrophism

Radical commentary (Gorz, 2010; Streeck, 2014; Žižek, 2010, 2014) asserts that a terminal crisis of capitalist modernity is no longer a prospect, it is breaking upon us. We moderns are forgiven for fearing that all is lost and that Jane Jacobs' (2004) dire prediction of epic recrudescence, a *Dark Age Ahead*, was terribly right. And yet this is surely not the final truth. We should have greater faith than our present despair allows in the capacity of modern civilisation to survive its greatest failures. Modernity released our species from an eon of grubbing, toil and servitude. There is no evidence that we humans have collectively given up on this great saga of emancipation. Indeed, its human appeal continues to widen in a long, relentless spread across the face of a cosmopolitan globe. It will survive if we do. And we will outlast the present crisis; battered and baked, to be sure, but our species will continue. Even the most appalling scientific predictions grant the Earth and its dependants some form of future. We may have to endure catastrophes, but the cause of catastrophism is wrong headed and certain to deflect us from the task of reclaiming the human prospect from our own follies. Humanity will emerge remade from the storms about to break, as will the planet we inhabit and shape, in ways we cannot yet foresee.

This is to assert the most fundamental understanding of resilience, as species survival – of course, raising the question of endurance, if not flourishing, for the non-human world. Still, the sirens of catastrophism must be resisted if resilience is to be more than Darwinian survivalism. The catastrophism of some radical (or at least marginal) perspectives would join cause with reaction by asserting an end to human history, certainly its modern phase. As Barry (2012, page 290) makes clear: 'a view of the inevitability of collapse can and does lead to de-politicized or even anti-political responses'. Predicting a post-industrial dark age for humanity, Kunstler (2005) sees history superannuated to *The Long Emergency*; centuries of starvation, disease and civil disorder (also Kunstler, 2012). In this schema, we moderns will return to 'reenchanted lives'. Both conservative reaction and progressive catastrophism refuse in different ways the work of species transformation which lies before us. We should not fear terminal social crisis, because, like death itself, it is part of the human condition, the necessary prelude to the rebirth of prospect. Arendt (1998 [1958], page 96) observed of *The Human Condition*:

> Life is a process that everywhere uses up durability, wears it down, makes it disappear, until eventually dead matter, the result of small, single, cyclical, life processes, returns to the over-all gigantic circle of nature herself, where no beginning and no end exist and where all natural things swing in changeless, deathless repetition.

And so it is surely with social life. We are learning painfully that, contra the Promethean delusion, human society is fastened to the 'deathless repetition' of 'natural things'. As our great modern accomplishments have shown, we are not a passive element in that eternal movement of existence. For us, recurrence means co-evolution with the planetary order, which has brought lately the urban age. The city is simultaneously our principal act of realisation, and our greatest disturbance of the natural order. From this costly plateau of achievement we must conceive and commit to a more settled path of planetary evolution. The times are certainly dire enough that resilience thinking should now assert so forcefully in global scholarship and policy thinking (Fainstein, 2013). As Streeck (2016, pages 37–40) asserts: 'the new proliferation of resilience thinking perfectly reflects an "age of entropy" where species energy and strength, and thus prospects for social solidarity in the face of manifold threats, seem at desperately low ebb'.

For progressive thought, however, this sets in train an imperative for new conceptual thinking, new political imaginaries to light the path to safer shores. Harvey (2012, page xvi) puts it plainly: 'Our political task … is to imagine and reconstitute a totally different kind of city out of the disgusting mess of a globalizing, urbanizing capital run amok'. Davis (2010, page 45), so exercised as noted earlier by global endangerment, refuses the path of defensive thought, instead calling for new imaginaries that envisage more than survival: 'only a return to explicitly utopian thinking can clarify the minimal conditions for the preservation of human solidarity in face of convergent planetary crises'. One imaginary not much drawn upon in contemporary urban debate is that of religious thought. Within this complex, multivalent set of traditions that have variously accommodated and opposed modernity, there runs a long conversation about cities that invites critical engagement. At the very least, in a time of depleted thought and political animus, there is nothing to be lost by such a venture for urban conversations that are increasingly focused on human endurance (resilience) in a time of threat.

Beyond Simple Secularism

Western 'faith traditions'[1] have already had a lot to say about city life, some of which has deeply influenced the modern journey of urbanisation, especially the search for what has long been termed the 'good city'. The urban imagination of early industrialism was anchored by scriptural icons (Davison, 2016). The two opposing biblical motifs, Jerusalem and Babylon, held sway and counter sway over Victorian thought. From the sacred texts came two different city prospects – a 'New Jerusalem' where social and personal virtues were perfected contrasting with a 'New Babylon' that bowed to the idol of material gratification. Sometimes Babel was enlisted as symbolic critique, for example to describe the institutional chaos of Victorian London. (Disraeli was even more scathing, describing it as a 'modern Babylon'.) The concept of evil also left its sooty footprint on Victorian thought. In raising eyes to a 'New Jerusalem' of urban perfection, Blake famously intoned against the 'dark satanic mills' bequeathed by laissez-faire industrialism. Hunt's (2004) account of Victorian urbanisation, *Building Jerusalem*, reminds us just how serious these biblical constructs were to the technocrats, entrepreneurs and reformers who conceived, built and remade industrial cities.

Through this lens emerges more clearly the contributions of religiously inclined or engaged thinkers to the canons of modern social science. This includes scholars normally regarded as strict secularists but whose work on closer inspection reveals the mutually constitutive work – at least in some registers – of religious and materialistic thought. The theologically mindful works of Hannah Arendt and Julia Kristeva provide powerful examples. Arendt's (1998 [1958]) *The Human Condition* was written in the awful wake of depression, war and holocaust. As a secular European Jew who lived through the Holocaust and contributed mightily to post-war humanist enquiry, she is regarded as a primordial critical theorist. In a key moment of her exposition, Arendt, however, resisted a restrictive scientism to enlist the New Testament and its announcement of glad tidings ('unto us a child is born') to frame her key concept of 'natality' (of which, more soon).

Arendt's book enormously influenced Western social science, especially political philosophy. Its most striking conceptual offering was that of natality, the boundless possibility of human renewal that arises from our urge to interpret and recreate the world we find ourselves in. This is no Whiggish idea of human history driven by bright men with big ideas, but one that binds us back to nature, which locates at the core of our re-creations the great wheel of procreation to which we are fixed as natural beings. It rejects the various doctrines of predestination – in science, the naturalisms which see human affairs as law bound; and in some religious outlooks, the teleologies that reduce the future to a final day of judgement.

Arendt finds the primordial story of natality in the scriptures. Dolan (2004, page 606) explains:

> She characterises this ineradicable possibility as nothing less than 'the miracle that saves the world' from the ruin to which it is otherwise subject. The greatest symbol of this possibility – 'its most glorious and succinct expression', Arendt says – 'is the Christian Gospels' announcement of glad tidings' … It is this Christian figuration of the miraculous through the image of the newborn that gives Arendt the term 'natality'.

Kristeva, a psychoanalytic scholar, has written closely on Arendt's work, including its religiosity (Kristeva, 2001, 2010) – a curiosity surely driven by Kristeva's own interest in the 'shortcoming of the humanist discourse' (Kristeva, 2010, page 27).

Another critical voice influenced by religious thinking (albeit from quite a different perspective) is recorded in the 1970s writings of Ivan Illich (e.g. 1977), a priest-philosopher considered a maverick in both academic and religious worlds. His fire-breathing critiques of Western society were driven by a radical eschatology which attempted that most difficult work of imagination; of a world without capitalism and the excessive trappings of industrial modernity. Illich was prepared to invoke bad gods as well as good to illustrate his critique of modern capitalism. He saw us enslaved to Prometheus, the ancient Greek deity who symbolised the hubris and destructive overreach of humanity. From Illich's time, as the field of social ecology emerged, 'Promethean' became pejorative shorthand for material waste and scientific overreach. It was freighted with the idea of looming catastrophe, of modern humans crashing disastrously back to the earth they sought to scorn and transcend through technologically driven progress. Illich did not reject modernity or even technological progress, but warned us to scale back, to recommit to the ancient principle of 'self-limitation' that can still be observed within indigenous cultures. His alternative future was not post-modern but a 'convivial modernity' where technologies ('tools') were limited by the principles of human and natural sufficiency. The bicycle was his ideal convivial tool. Illich (who died in 1992) would surely be delighted to see its new centrality in Western cities.

As the work of these radical scholars shows, resort to religious thought and scriptural exegesis may or may not require faith, but it arguably opens the social scientist's mind to new realms of metaphor, reason and imagination. Perhaps, as with Illich, it helps us to think the unthinkable, to journey in thought beyond the confines of formal reason to other worldly possibilities. In these perilous times, when the future seems more terribly foreseeable than reassuringly imaginable, there seems compelling need to subject science (albeit with care) to the speculative world of faith. As Eagleton (2015: page 112) offers: 'Faith and hope are most needed where knowledge is hard to come by'. This is not to embrace the anti-scientific antipathies of fundamentalisms that thrive in endangered modernity (Beck, 2009). Rather recourse is urged to the richly conceived, open-hearted speculations about the 'good life' (and the good city) that are found in faith traditions, Christian and otherwise.

Arendt reminded us that doubt was a fundamental Enlightenment value, the restraining twin of reason. Without it, reason is unchecked – excessive, deadly rationalisation too often follows (think Nazism). The late Ulrich Beck (1993) urged us moderns to recultivate 'The art of doubt' so that science and reason could once again be tied to the cause of human improvement, not venal power or planetary consumption. Faith too, as Eagleton (2015) explains, can be strengthened not undermined by doubt, most especially the sceptical interrogations of science. Pope Francis confirms: 'In this quest to seek and find God in all things, there is still an area of uncertainty. There must be. If a person says that he met God with total certainty and is not touched by a margin of uncertainty, then this is not good' (Spadaro, 2013). In a new mutually restraining play, faith and reason might help to clarify a human imagination clouded by the whirling disenchantments and fears of a failing modernity.

Seeking the Good (Enough) City

A recent exposition, *Seeking the good enough city*, by the US-based theologian Bretherton (2015) is suggestive of how dialogue between religious and secular traditions might enrich rather than simply antagonise human discussion in an epoch of crisis. His argument enlists secular theory as well as theology and scripture to arrive at a surprisingly pragmatic casting of the good enough city. Bretherton's paper opens by considering the contemporary urban preponderance and its many inadequacies and contradictions before proceeding to outline the case for a 'good enough' city that would meet the expectations of Christian humanism.[2] (His position is expressly at odds with religious fundamentalism.)

A first point of convergence between the secular and the spiritual in Bretherton's account is the disavowal of earthly urban utopia. Both traditions are capable of sharing, if uneasily, a common ground on this critical point. The secular left, especially so. As Eagleton (2015: page 40) puts it: 'Images of utopia are always in danger of confiscating the energies that might otherwise be invested in its construction'. Arguably, utopian thinking about cities has rarely produced anything other than misty-eyed simplisms that have failed to engender transformative change.

There are other elements of Bretherton's account of human purpose in the urban age which can be interpreted, and to some extent supported, by critical social science. The first is the sanction of exile that is simultaneously the story of the Christian people and, arguably, the historical motif of modernity. Bretherton finds the story of exile in Augustine's 5th-century treatise *De civitate Dei* (The city of God) which itself draws upon the writings of Jeremiah, a prophet recognised by all three major monotheistic religions. Jeremiah 29 states that it is the will of the Lord of Hosts, the God of Israel, that the chosen people will accept banishment to Babylon, the secular city, but as a place of flourishing not deprivation; as a place to be good in, including accepting of authority. For the chosen, the secular city is not to be a place of disconsolation and refusal but a place where (enough) goodness can be found. Jeremiah: 'seek the welfare of the city where I have sent you into exile, and pray to the Lord on its behalf, for in its welfare you will find your welfare'.

Augustine's *De civitate Dei* relieves the metaphor of exile a little by describing the people of God as pilgrims residing in Babylon, enjoying its peace and abundance whilst quietly refusing its worldly failings through the witness of faith. Augustine enjoins both the faithful and the unbelievers to share and maintain 'the temporal peace' of Babylon. This resonates with the long journey of urbanisation in modernity where the city came to symbolise (if not always provide) the virtues of justice, peace and freedom. The early influxes to Europe's nascent cities were from 'feudal pilgrims' seeking relief from bond and toil in the countryside. 'Stadtluft macht frei' was the promise of the times. The project of modernisation is a great species pilgrimage that seeks improvement not utopia. As Erich Fromm (2009 [1942]) reminded us, it is a fraught journey towards freedom that imposes new burdens, especially the insecurity and anomie of modern urban life.

This brings us to the urban question. If nothing else, modernity has been a great recamping of our species, a journey from rural to urban, from enchantment to reason, from servility to freedom. On the face of it, a great liberation from 'the idiocy of rural life' as Marx and Engels unkindly put it (Marx and Engels, 1985 [1848]). However, digging deeper – as indeed Marx did with the greatest insight – it was at the same time a monstrous work of separation, from natural caprice yes, but also from nature itself. In the left tradition, this story was put into verse by Bertolt Brecht:

> I, Bertolt Brecht, came out of the black forests.
> My mother moved me into the cities as I lay
> Inside her body. And the coldness of the forests
> Will be inside me till my dying day.

In Brecht, the prospect of complete severance is rejected. We pilgrims from the forests will carry Nature 'inside us' to our dying days. The dual weights of modern life are this species freight, our nature and what Fromm (2009 [1942]) called 'the terrible burden of self-strength'. The record shows that we have struggled to recognise let alone sustain the yokes of modern existence. Freud attempted to identify and explain the prices of civilisation, which humans accumulated from their earliest life experiences. He asserted that we cannot construct a future when held by the undertow of the past. (He also dismissed religion as an illusion, but wrote much about it.) Modern Anglophone psychology exhibits little interest in his thesis. The now preponderant cognitive behaviour therapy is transfixed with the integrity of the present, perhaps evoking a civilisation that cannot mobilise the imaginative energy that is needed to face an imperilled future (compare Kristeva, 1989, pages 221–2).

Surely the greatest threat to any pilgrimage is that of alienation – to be trapped in exile from human meaning and possibility. Bretherton's acceptance of this notion represented a second convergence of faith and reason, at least in his account of urbanism in troubled times. Alienation is the corruption of modern possibility that was central to Marx's critique of capitalism. It stands like a dark wall in the way of human realisation. Arendt warned of the consequences of modernist alienation, which might be summarised as an outbreak of human stupidity. The 'almost infallible signs of alienation from the world' would be, 'A noticeable decrease in common sense in any given community and a noticeable increase in superstition and gullibility' (Arendt, 1998 [1958], page 209). The waning of human common sense is perhaps powerfully registered in climate scepticisms and the misty-eyed enthusiasms for pre-modernity that have declared a 'New Age'. Both, in very different ways, are 'denialist' movements, retreating into mysticism in the face of awful human endangerment.

Bretherton confronts modern urban alienation and a growing impulse for secession amongst the socially disenchanted. He sees this starkly represented in the contemporary gated community. His thesis decries the alienation imposed by urban segregation, asserting that all citizens must accept that they are 'participants in a shared community of fate' (Bretherton, 2015, page 2). For him, the duty of social belonging is scripturally sanctioned. It surely follows from Jeremiah and later Augustine, where the pilgrims must join themselves to the common Babylonian purpose, and by consequence, the inevitability of shared fate. Civic refusal is not licensed by faith. This is of course at odds with many current and historical faith outlooks that have ignored Jeremiah's injunction through various forms of bunkering and self-imposed exile. The separatists take Paul's injunction in Romans 12:2 ('Do not be conformed to this world') to an extremity that seems to deny any faith in common human virtue. This is of course a position that deeply rejects any defensive notions of resilience.

The notion of shared fate is also confirmed in secular reason where denial of species being, of sociality, conflicts with any rational understanding of the human condition. This assumption has been central to most accounts in Western social thought. Bretherton draws from one of its contemporary streams, communitarian theory, finding the path towards goodness – never to be realised, but always to be aspired to – in democratic politics, avowing a pluralistic model that privileges community organising.

The engagement of secular and religious humanisms in Bretherton's account of the 'good enough' city suggests one form of response to Harvey's call 'to imagine and reconstitute a totally different kind of city' (Harvey, 2012, page xvi) in perilous times. However, divergences are also apparent, especially between the basic postulates and prescriptions of critical materialism and Christian humanism. A wholesale regeneration of local democracy will be necessary to achieving a 'good enough city', but it won't be enough in the first instance to generate the new human dispensation that must replace neoliberal capitalism, perhaps in the smoking ruins of what it leaves behind. Bretherton's account needs filling out to better engage the asymmetries of power (which he recognises as a problem) that plague all striving for better purpose in a failing world order. To state the point bluntly, as Harvey (2012) has done, what hope has localised city activism in the face of globalised capital that sees urban development as a treasury of super-profit?

Conclusion: Resilience without Hope

In an age of depleted progressive thought and looming urban endangerment, we should exhibit more sensitivity to the complex and seemingly antithetical traditions from which we draw critical and creative insight. This includes religious thinking and surely indigenous knowledge. We will need these resources joined as counterforce to the dangerously entropic force of post-ideological capitalism. This means, of course, ruling out the dangerous fundamentalisms that Beck (2009) explained as the flowers of evil spawned by a dying modern order. Reason can look to the wide-eyed humanisms of all faith traditions as sources of new imaginative energy and insight. Their generally insistent interest in human ethics, values and conduct can enlarge and enrich secular conversations about alternatives to the present (compare Kristeva, 2010, pages 209–12).

Resilience is a hard urban shore – philosophic, religious and traditional thought reminds us that we can be more than sentient, rational beings, that we can strive for and enjoy goodness. It also reminds us that this is a journey – a pilgrimage – never completed. Fainstein (2013, 2015) is right to criticise biologic renderings of resilience that rely on naturalistic, mechanical notions of evolution. But perhaps evolution can be enlisted more as humanistic metaphor to describe a transhistoric species journey that must inevitably encounter and embrace the imperative of transformation. This raises the question of how to live in every new state that our pilgrim species will find itself in, including whatever urban dispensation lies beyond the perils of the present. We may well attain a good enough city as a new human heartland, but how will we live together in it? As Bruce Springsteen anguished, 'It's hard to be good in the city'. Perhaps, as Fainstein suggests, there is no better starting point than 'the need of poor or vulnerable people' (Fainstein, 2013, page 10) – that is, the deep end of social resilience. We need to subscribe all philosophies of the good to the task of creating a new urban world.

Eagleton (2015) argues that hope is the ideal that must bind any search for a new human dispensation. It sounds obvious, but it isn't, because hope must first be shorn of optimism – an unreasoned faith in the future loaded with the deadening influences of utopianism and ... well, simple-headedness. Eagleton's appeal to religious as well as secular thinking is intentional as he sees both traditions offering conceptions of hope that in different ways look like 'cautious pessimism' – put differently, critical realism. This resounds with Streeck's (2016) recent criticism of the tropes of resilience and hope in contemporary 'end times' capitalism, where the increasing popular currency of both indicates their negation, and a desire amongst elites, including policy elites, to rhetorically deny or at least defray planetary scale peril.

In our hour of species peril, the hopeful seek a new world but only from the ruin that seems sure to come, ordained by an order stubbornly clinging to its destructive prowess. Retrenchment of capitalism will not present a straight path back to ecological moderation or social peace. Apart from sparking heaven knows how many more gruesome wars and migrations, its death agonies will likely generate many desperate quests for salvation through resource exploitation. These misadventures are prefigured in the contemporary lust for Arctic exploitation, the fracking rush in the new worlds, and the enthusiasm for newly unlocked carbon, such as Canada's tar sands. Prometheus it seems will rein until dethroned by collapse – all cries for abdication will go unheard. The task seems no longer to impose a better future but to prepare for an imposed future.

Capitalism may well have entered its terminal phase, as the late Andre Gorz (2010) firmly believed, but this does not mean the end of humanity or even modernity. Our species will have to take to the road of history again. Looking ahead to prospect, Eagleton (2015, page 115) writes: 'Hope ... is what survives the general ruin'. Further: 'Though there will be no utopia, in the sense of a world purged of discord and dissatisfaction, it is sober realism to believe that our condition could be mightily improved. It is not that all will be well, but that all might be well enough' (Eagleton, 2015, page 133). For the hopeful – the faithful and the reasonable – the ending of the current world order means a new journey not a termination of history. It represents – to join Bretherton's (2015) narrative – another stage in the human pilgrimage, to a new Babylon, where the search for the 'good enough' city must be renewed.

The close of Ulrich Beck's 1993 book, *The Reinvention of Politics*, ponders the imponderable, a modernity that has run out of ideas, exhausted by hubris and depleted by failures. And yet, he doesn't lose faith in the project, seeing portent in ruin. The last sentence reads: 'Only a final lack of options frees oneself, but you still hope and you're dangerous' (Beck, 1993, page 177). Perhaps the most dangerous and hopeful idea in the present is to not fear catastrophe, seeing it instead as a means to transformation and renewal. To enlist Fainstein (2013), an evolutionary step forward that prioritises the needs of the vulnerable – in other words, to recommit to the necessary but ever confounded journey of justice. Here the poor and the vulnerable cannot be dismissed as ghosts in a resilience machine, but privileged as the first subjects of human renewal.

If we take Arendt seriously, we should not fear terminal social crisis because, like death itself, it is part of the human condition, the necessary prelude to rebirth of prospect. Having closely observed that frightful 20th-century trinity, depression, genocide and world war, she surely knew what she was talking about. The miracle of species recovery that awaits us is, as Arendt explained, our inborn ability to endlessly produce the new ... against the odds. And in the ashes that are left to hope, we must have faith that the child will be reborn. In the hour of species peril, is this what resilience can mean?

Notes

1 Surely other traditions too, but I own no expertise in them.
2 A framework as I understand it informed both by scripture and humanist philosophy.

References

Arendt, H. (1998 [1958]). *The human condition*. Chicago, IL: University of Chicago Press.
Barry, J. (2012). *The politics of actually existing unsustainability*. Oxford: Oxford University Press.
Beck, U. (1993). *The reinvention of politics*. Cambridge: Polity.
Beck, U. (2009). *World at risk*. Cambridge: Polity.
Beck, U. (2012). *Twenty observations*. Cambridge: Polity.
Bretherton, L. (2015). 'Seeking the good enough city'. Paper presented to forum: What makes a good life for a city? 30 September. Melbourne Cathedral series, University of Melbourne.
Brugmann, J. (2009). *Welcome to the urban revolution*. St. Lucia: University of Queensland Press.
Davis, M. (2006). *Planet of slums*. London: Verso.
Davis, M. (2010). 'Who will build the ark?'. *New Left Review*, 61: 29–46.
Davison, G. (2016). *City dreamers: the urban imagination in Australia*. Sydney: University of New South Wales Press.
Davoudi, S. (2012). 'Resilience: a bridging concept or a dead end?'. *Planning Theory and Practice*, 13(2): 299–333.
Davoudi, S. (2016). 'Resilience and the governmentality of unknowns', in M. Bevir (ed.), *Governmentality after neoliberalism*. New York: Routledge, pp. 210–249.
Dolan, F.M. (2004). 'An ambiguous citation in Hannah Arendt's *The Urban Condition*'. *Journal of Politics*, 66(2): 606–610.
Eagleton, T. (2015). *Optimism without hope*. New Haven, CT: Yale University Press.
Fainstein, S. (2013). 'Resilience and justice'. Research paper no. 2, Melbourne Sustainable Society Institute, University of Melbourne.
Fainstein, S. (2015). 'Resilience and justice'. *International Journal of Urban and Regional Research*, 39(1): 157–167.
Fromm, E. (2009 [1942]). *The fear of freedom*. London: Routledge.
Glaeser, E. (2011). *The triumph of the city: how our greatest invention makes us richer, smarter, greener, healthier, and happier*. Harmondsworth: Penguin.
Gleeson, B.J. (2014). *The urban condition*. London: Routledge.
Gorz, A. (2010). *Ecologica*. London: Seagull Books.

Harvey, D. (2010). *The enigma of capital*. London: Profile Books.
Harvey, D. (2012). *Rebel cities*. London: Verso.
Hollis, L. (2013). *Cities are good for you: the genius of the metropolis*. London: Bloomsbury Press.
Hunt, T. (2004). *Building Jerusalem: the rise and fall of the Victorian city*. New York: Owl Books.
Illich, I. (1977). *Toward a history of needs*. New York: Pantheon Books.
Jacobs, J. (2004). *Dark age ahead*. New York: Random House.
Kahn, M.E. (2010). *Climatopolis*. New York: Basic Books.
Kasarda, J. and Lindsay, G. (2011). *Aerotropolis: the way we'll live next*. London: Allen Lane.
Kristeva, J. (1989). *The black sun: depression and melancholia*. New York: Columbia University Press.
Kristeva, J. (2001). *Hannah Arendt: life is a narrative*. Toronto: University of Toronto Press.
Kristeva, J. (2010). *Hatred and forgiveness*. New York: Columbia University Press.
Kunstler, H. (2005). *The long emergency: surviving the converging catastrophes of the twenty-first century*. New York: Grove/Atlantic.
Kunstler, H. (2012). *Too much magic: wishful thinking, technology, and the fate of the nation*. New York: Atlantic Monthly Press.
Marx, K. and Engels, F. (1985 [1848]). *The communist manifesto*. Harmondsworth: Penguin.
OECD (2010). *Cities and climate change*. Paris: OECD Publishing.
Resilient Melbourne (2016). *Strategy*. Available at: http://resilientmelbourne.com.au/strategy/
Sayer, A. (2011). *Why things matter to people: social science, values and ethical life*. Cambridge: Cambridge University Press.
Spadaro, A. (2013). 'A big heart open to God'. *America. The Jesuit Review*, 30 September. Available at: http://americamagazine.org/pope-interview
State of Victoria (2017). *Plan Melbourne: metropolitan planning strategy*. Available at: www.planmelbourne.vic.gov.au/Plan-Melbourne
Streeck, W. (2014). 'How will capitalism end?'. *New Left Review*, 87: 35–64.
Streeck, W. (2016). *How will capitalism end?* London: Verso.
United NationsHabitat (2012). *State of the world's cities 2012/2013*. UN Human Settlements Programme, Nairobi, Kenya.
World Bank (2010). 'Cities and climate change: an urgent agenda'. Urban development series; knowledge papers, 10. Washington, DC: World Bank.
Žižek, S. (2010). *Living in the end times*. London: Verso.
Žižek, S. (2014). *Trouble in paradise*. London: Allen Lane.

12

DISMANTLING THE RESILIENCE MACHINE AS A RESTORATION ENGINE

Timothy W. Luke

This study is a preliminary exploration of currents of thought and practice associated with the idea of resilience in contemporary politics of neoliberal economies and societies at dusk for the Holocene epoch. Consequently, it addresses a growing concern, namely, what is the present state of humanity and non-humanity on the Earth in times that are increasingly described as catastrophic rather than normal? How might individuals and groups respond, react or recover when ordinary policy conditions, once thought to be stable, are unexpectedly disrupted on a catastrophic scale? And why are particular cycles of action/reaction defined as resilience and becoming dominant themes in many neoliberal policy debates?

This exercise investigates the ontopolitical drift in neoliberal policy and politics at the advent of what some claim is a new geological epoch, namely, the Anthropocene and the end of the Holocene (Radkau, 2008). This possible shift is significant because the idea of resilience, unlike stability, may imply a rising risk of greater instability as well as a general proliferation of widespread degradation, destruction or disruption to which policy making continuously must respond.

During the Cold War era, the policy-making process faithfully upheld goals of state-regulated stability, assuming that catastrophic destruction in nuclear war or economic crisis was possible but not too likely as long as policy makers maintained their vigilance. If instabilities did develop, many believed effective state-policy solutions were always ready to re-establish stability. Indeed, these policy responses could quickly be mobilised, and the semblance of normality would be maintained. With neoliberal economic and social reforms during the 1980s and 1990s, however, the regulative firewalls of state oversight were breached, and many Cold War geopolitical tensions also eased. In turn, rapid economic growth ignited during the 1990s in many places around the world.

Yet as this economic growth decelerated in 'the dot com bust' of 2000, and then failed in many places during the Great Recession of 2006–9, confidence in state regulation and centralised stability-enforcing mechanisms waned.

Political security and economic stability fragmented, at the same time, into seemingly permanent economic instability, social turmoil and institutional gridlock. After neoliberal restructuring came a new set of hard realities about public policy. That is, persistent failure rather than recurring success can quickly take hold and persist as a very common outcome, which most consumers and producers do not easily acknowledge. Policy theory and practice, in turn, began to highlight the promise of resilience, or 'bouncing back' from major disruptive reversals in this turbulent new time, to foster hopes that normality and stability could be 'restored'. With resilient individual attitudes and group associations, something like what had held true prior to a catastrophic turn of events could eventually return. For those seeking to sustain development, attain democracy or maintain decency, faith in this restoration engine spinning at the core of new resilience practices should not be ignored (Walker and Salt, 2006).

Are Anthropocene scenarios of life on Earth, however, a historic break? Proponents of the Anthropocene concept have proclaimed that global society now is living through the 'Great Acceleration' (McNeill and Engelke, 2014; Luke, 2015, pages 139–62). Yet, this complex mystification mainly documents how a few great powers and wealthier countries, mostly among the now G20 bloc of more 'Westernised' nations in different regions of the planet – which accounts for 80 percent of world population, 80 percent of world trade and nearly 85 percent of the gross planetary economic product – are accelerating, while the powerless and less great of 'the Rest' are being left behind, or even run over, in great numbers year after year. When speculating about these massive changes, Anthropocene-leaning thinkers tend to judge very superficially from recent trends, not long-term confirmed scientific realities (Ehlers and Krafft, 2010). With their taken-for-granted confidence in human rationality's capacity to discern the origin and import of certain indisputable geophysical/cosmological/existential truths accurately, policy experts basically offer to all what they regard as state-of-the-art technical solutions for 'resilience science' (Resilience Alliance, 2010). Ultimately, these cadres of experts hold out the hope of resilience discourse to those who would endure. Assuming destruction is normal, and knowing disruption will be unavoidable, the populations of every country are encouraged by such experts to accept fully the directives of resilience so that they will be able and ready 'to build capacity', 'maintain function' and 'absorb disturbance' (Walker and Salt, 2012).

Resilience Narratives and Catastrophism

Following the lead of actuarial science, disaster planning or security preparedness studies, decision makers and scholars have, as the Holocene wanes, become taken with 'catastrophism' with respect to destructive episodes of disruption that either

'Humanity' is wreaking on the 'Environment' or the 'Environment' is imposing on 'Humanity' (Biro, 2015, pages 15–37). This reciprocal impact, in turn, is used to justify a major existential shift into contemporary biopolitical practices and techno-scientific thinking, namely, to demand that groups and individuals must let go of stability, security and safety. Like soldiers, who often live only by accepting they already are, will be or should be dead simply have a chance to return home alive, civilians also now are told that they too must live out their everyday lives aware of endless intense endangerment. In accepting the slow violence, inevitable disruption or catastrophic impact of economic collapse, global terrorism and climate change that are existentially embedded deep disasters in contemporary life, everyone is required to gain resiliency if any normality is ever to be restored.

Living as if life soon will shatter, and then preparing to return it to some status quo ante as robustly as possible, is an elaborate fresh mythography for everyday life evolving within the discourses of 'resilience'. It has become the strategic counterbalance in the public debates about the world's endemic instability and rising chaos. Resilience basically means springing back, bouncing back or settling back into some prior state. Yet, it also can denote springing away wildly, bouncing around erratically or rolling away aimlessly. If the prior conditions are already plagued by disarray and destruction, there are usually harsher truths hidden in such recoil cycles. That is, crisis encounters indeed might entail springing off into disarray, bouncing far off track or running totally off the rails of normality.

Often, at best, the recoil of bouncing back only kicks the resilient back to those conditions of misery, instability or stagnation that prevailed prior to the massive disruptions that then forced them to claw back to those abject conditions. Genuinely complete restorative revitalisation is rare. Instead, the restoration engine that is allegedly ever ready to get everyone back to where they dream of returning does not start. Instead, the resilience mechanisms for groups and individuals repeatedly blow back the bolts of their brutal degradation into the cold, hard receivers of misery in which the firing cycles of continuous destruction recycle rather than readjust. This hidden containment in resilient recoil carries a hard kick, and it frequently will only return the disrupted back to the lesser miseries they endured prior to their latest disastrous disruptions.

Like the urban growth machine, the resilience machine valorises a known unknown, 'resilience', and then forges a routinised set of functionalistic responses for restoration, which perpetuate their own responsive mentalities, instrumental objectives and machinic perspectives. Just as all cities tend to believe all growth is good, so, too, do resilience networks affirm that cultivating more individual and collective resilience is always imperative. In many contexts, these parallels become co-dependent alliances. When 'growth' stalls, fails or ends, 'resilience' is pushed as the answer. While the range of 'resilience' practices has grown in scope and number, there are no guarantees they can provide immediate relief. Yet, they are readily believed to trigger resilient spurts of 'growth'.

Trust in this simple saga of having normality, losing normality abruptly and then seeing normality restored via resilient preparation begins elsewhere. Following the trajectory of many scientific ideas, resilience has been lifted from other disciplines, ranging from ecology to psychology, and placed where it perhaps ought not to be. Clearly, the dense loops of resilience theory and practice can attract many to accept hard kicks of vast losses in the misplaced faith that restorative gains will follow. Still, they are not inevitable, and all too often such gains only renew fixed routines that lock and load losses more sustainably rather than truly regain what remakes things right. This popular understanding of resilience seems attractive, but its workings hide the hard recoil pounding within such constrained restorative cycles.

Holling's analyses of cyclical evolutionary development in ecosystems, for example, are germane. He depicts resilience ecologically in lean narratives of systemic growth, then collapse and finally reorganisation (Holling, 1978). System reorganisation is not certain, but it will spark new growth or stall in entropy, disorganisation or catastrophic collapse (Davoudi, 2014). Resilience, therefore, equals a system capable of undergoing all of these changes without losing its distinctive attributes of survival. This tacitly normative, but avowedly objective, concept is very alluring for those experiencing ecological and economic systems in crisis (Gunderson and Holling, 2002). Nonetheless, major influential texts about resilience ignore the hidden recoil in resilience, and stay solidly on this normalised direct line of march (Walker and Salt 2006).

Resilience and the Anthropocene

The recent intervention by numerous scientific and technical authorities, which monitor the unfolding disasters of the 'Anthropocene', appears to naturalise the harsh and hidden recoil of resilience. In fact, the naturalisation of anthropogenic disasters is driven deep into geological time as part of the current generation's fascinating acceleration in historical time during the Cold War era's ethical, political and social mobilisation (Crutzen and Stoermer, 2000). On one side of the ledger, resilience constitutes accepting a permanent state of emergency aimed at building capacity in neoliberal capitalism worldwide for people 'to do something' about all possible forms of destruction, while, on the other side of the ledger, enlisting them in such emergency exercises occludes how Nature itself is being reshaped as a growing anthropogenic disaster that now recharges these recoil cycles over and over.

In the sweeping mass depoliticisation of global economic stagnation, various networks of resilience science and adaptive technology experts are arguing that 250-plus years of economic expansion tied to intensive fossil fuel use equal both 'the rise of Western civilisation' and 'the end of Nature' as world-historical catastrophes that will continue compounding for decades. How to manage the planet's endangered populations from above and afar amidst this catastrophe is the contemporary challenge. Can any collective effort restore them and their milieux to how and why they were prior to the disruptive anthropogenic changes in the Earth's climate patterns during the 1760s? The unrelenting patterns in humanity's growing civilisational footprint for over 250 years clearly constitutes a circular trail of attempted recoveries interspersed with bursts of unchecked growth that do not lead back to reclaiming normality by restoring the pre-industrial climate. It leads instead to repeatedly recoiling back into more varied and extreme states of destructive disorder.

Even though returning the Earth to pre-industrial levels of CO_2 in the atmosphere would take decades upon decades to occur, the mythographies of that restorative possibility are naively confirmed by scientific communities as well as lay publics as a 'can-do' goal to anchor their efforts and focus their attention. Resilience, as the work of Paul Crutzen illustrates, is a condition of being, and it pivots upon drawing our awareness to a planned package for rebecoming the same, if not better. Yet the consequences of our fossil fuel-burning past and present, with the gigatons of new greenhouse gas emissions being vented every day, necessarily will stall these outcomes regardless of whatever future collective actions are attained at great speed and possible permanence. Many people want to pretend restoration is attainable, but there are no guarantees. 'What I hope,' Crutzen says, 'is that the term "Anthropocene" will be a warning to the world' (Kolbert, 2013, page 33). Granted his ontographic warning about today's social and natural reality, few heed his warnings on face value. Pushing to operationalise his resistance to greenhouse gassing, as their resilient bounce-back, is not fruitless, but most ignore such recoil pain.

Resilience thinking in an age of considerable crisis and continuous disturbance arguably is this historical moment's efforts to express its enlightenment, and it thereby constitutes a critical ontology of ourselves. Like work, being resilient is repeatedly mobilised to describe, if not define, individual and collective effectiveness in times of turmoil. If one works hard, he or she must succeed. Should one fail, it is evident he or she did not work hard. By the same token, in times of trouble, the resilient are those who rapidly rebound from shocks, and then recover from any disturbances as resiliency demands. Those who never seem to get over crises always are slow to rebound or usually never seem to recover. Thus, it also becomes evident they are at fault for not building resilience. Consequently, resilience usually is not necessarily attaining a restorative gain as much as it is withstanding the punishment of embedded pounding in resiliency's recoil.

In these times of economic chaos, endless low-intensity wars and embedded technological change, everyday life itself is shocking, crisis-ridden and always disturbed. Training people to live in neoliberal capitalist markets by embracing the raw, hidden recoil of resilience comes to define the ideal subject who will best adapt to these times (Evans and Reid, 2014). Meanwhile, the subject who lacks resilient recuperative capacities is framed ideologically by his or her inefficient, slow or never-attained rebound from the repeatedly worsening recoil. Here, speed at the reset, the scope of quick rebound and the solidity of recuperative responses are represented as the true affirmations of resilience and/or the tawdry displays of an insufficiently resilient life.

We ought not to ignore the history behind the resilience idea's problematisation, and one must ask tough questions about the ground for turning to such concepts. At this juncture in the Holocene, one must wonder, 'why a problem and why such a kind of problem, why a certain way of problematizing appears at a given point in time' (Foucault, 2007, page 141). By vindicating neoliberal economic instability, geopolitical insecurity or the Anthropocene epoch as valid pragmatic benchmarks, the agitation of resilience-science advocates leads one to reconsider 'the way in which things', such as the expressions of resilience in biopolitical strategies of governance, 'become a problem' (Foucault, 2007, page 141).

What has been unfolding in the normal science of resilience analysts, sustainability-minded environmentalists and global-complexity modellers is the recognition that all prevailing theories and professional practices have reached 'a point where in some way the certainties all mix together, the lights go out, night falls, people begin to realize that they act blindly and consequently a new light is necessary, new lighting and new rules of behavior are needed' (Foucault, 2007, page 15). By the same token, resilience science and its policy proponents struggle to produce answers: is this new light what is needed, are its beams that much brighter, can it bring fresh illumination on the challenges at hand and will it lead people to new rules of behaviour that truly are an improvement?

In this respect, the invention and popularisation of resilience rhetoric must be contextualised, conceptually and discursively, as 'the history of actuality in the process of taking shape' (Foucault, 2007, page 137). Strangely enough, policy and scientific debates about resilience are now another measure of how engaged every endangered life unfolding in this world becomes in constituting 'a history of the present: what are we and what are we today', namely, coping with the raw recoil of being unsafe, insecure and unstable (pages 16–137).

The Restoration Engine in Systems of Systems

The avowed purpose of resilience 'is the need to see the world as consisting of a larger number of different systems, small and large, natural and physical, often combined in complex ways – on which we depend' (Walker and Salt, 2012, pages ix–x). This openly declared epistemic, ontic and pragmatic set of assumptions to apprehend the world as a system of systems, working on different scales and in varied modes of complex articulation, is a highly motivating methodological policy construct. Fortunately, its structural simplicity provides a pretext for its cynical management in neoliberalism. As imaginative schematics to mobilise energy, information and resources, they also presume publics and their leaders will co-create policy programmes for the administration of normal systemic dynamics (if and when they are restored) as well as responses to abnormal systemic disruptions (since by and large they are the new normal).

Plainly, complex systems, as resilience science understands them, are found at multiple levels of analysis and many layers of genera. The promiscuity of resilience as a buzzword resonates in many fields of study, ranging from information networks, financial exchanges and electrical grids to corporate operations, governmental agencies and military services. For this discussion, however, one must concentrate upon 'the resilience of social-ecological systems (linked systems of humans and nature)' (Walker and Salt, 2012, page 1). The well-known issues of sustainable use, capacity overshoot, biodiversity loss and population growth are now directly plugged into every concern related to rapid climate change. The threshold of hidden recoil, then, rises as new conjunctures are generating hitherto unanticipated and/or unintended risks of major import to individuals and collectives dependent upon complex coupled social-ecological systems.

The baseline systemic parameters of robustness and fragility, carrying capacity and obvious overshoot, static stability and dynamic stagnation typically are gauged by most studies against the Earth's most recent geophysical era, the Holocene. This benchmark of terrestrial normality has come into question as the Anthropocene has been popularised (Marsh, 1965; Roberts, 1998; Ruddiman, 2005). Whether this shift is transpiring, though, is an interesting question, since resilience thinking focused until quite recently only on social-ecological systems as global systems. In turn, many resilience experts only examine the systems per se for those recurrent shocks they are enduring now. The product of these shocks might be creating a new geological epoch in the long run; but, in the short run, no matter what epoch the stratigraphers conclude the Earth is entering or leaving, resilience thinkers ask another question.

When managing social-ecological systems in times of 'climate catastrophes and natural disasters', resilience managers need to know 'how much can these systems take and still deliver the things we want from them', which remains 'in a nutshell' what is truly 'the central question behind resilience thinking'. Answering this question, administering probes to gauge overall system limits for getting that answer, and applying to maintain the delivery scheme for wanted outcomes correctly is how resilience thinking 'adds value to the way we manage the systems around us' (Walker and Salt, 2012, page x). These interventions also turn resilience thinking into a distributed technics for legitimating 'the restoration engine' as what can apply, justify and sustain the policy pragmatics of systemic overshoot mobilised at this juncture in the history of the Earth's societies and ecologies.

For many, resilience is a question, in part, about how to live without fossil fuels, since the concentration of 'carbon dioxide in the atmosphere has jumped 41 percent since the Industrial Revolution began in the 18th-century' (Gillis, 2013, page A4), while current fossil fuel consumption daily intensifies this destructive trend. Soon the effects of rapid climate change will be more pronounced than those to date – additional intense heat waves, extreme precipitation events, altered growing seasons and flooded coastal regions. As Jeremy Shakan, a Harvard research scholar, observes: 'We're just entering a new era in Earth's history. It will be an unrecognizable new planet in the future' (Gillis, 2013, page A4; and Davoudi, 2016, pages 152–71). The Holocene's conventional knowns are rapidly slipping into the radical unknowns of the Anthropocene.

Resilience as mere adaptation to the 'Great Acceleration' harmonises, as Marzec (2012, page 40) claims, with the aspiration to accomplish attaining a new condition of complete environmentality to reappraise, and then recuperate from, these brutal conditions of ecological unrecognisability on the planet Earth. Because of rapid climate change and other environmental threats, this environmentality becomes 'a political constitution and administration of planetary biosystems on the basis of the ability to be technologically "improved" so as to produce "high yields" of both consumption (more crops to feed people) or more energy for fuel and electricity' (page 40). Unfolding in parallel with total telluric environmentality, the challenges presented by rapid sea rise, weather disruptions and agricultural failure underscore how the environmentality 'should be understood precisely as being systemic in nature, not occasional, and not the result of an individual's or group's actions' (page 42). Instead, systemic management of social-natural systems is itself believed to be, as Žižek asserts, 'objective, systemic and anonymous' (2000). Resilience science occludes the hidden recoil of its restoration engine by talking about the easy terms of adaptive and mitigating theory and practice, which aim at developing 'all new theoretical categories through which to investigate the relentless production and transformation of socio-spatial organization across scales and territories' (Brenner and Schmid, 2011).

In this respect, Walker and Salt concur with Foucault on the power/knowledge nexus in that neoliberal practices always 'need to be in a "resilience frame of mind" to begin with' (Walker and Salt, 2012, page 1). Knowing how to identify what must be known gives resilience thinkers the power to tag the systemic thresholds, domains of action/interconnection/impact and range of adaptive cycles in planetary management and thereby specify how far systems actually are from thresholds of collapse as well as generally try to manage disturbances with regard to their tumultuous effects. Resilience practices always require 'a dynamic and adaptive approach', but the main objectives still are only monitoring 'the system's capacity to manage a disturbance and prevent the state of the system from reaching a threshold' (page 25).

Thresholds basically 'define the "safe operating space" of systems, and those effects are important inasmuch as they balance variables of concern and control' (Walker and Salt, 2012, page 8). Thresholds are complicated conjunctures of variables, but they basically boil down to: a) no effect; b) a step change; c) an alternate stable state; and d) an irreversible change. Efforts by social-ecological system managers to calibrate these shifts against a set of mapped and measured coordinates in the lost ecological spatialities of 1990, 1970 or 1950 are likely to fail, even if there were trustworthy tools for policy making to use for steering toward that return. What is not fully understood as society, ecology or a system cannot be managed, and little of it is found elsewhere to become more discernible (Ehlers and Krafft, 2010).

How and why the idea of the Great Acceleration and the Anthropocene are circulating in many scientific communities, such as atmospheric chemistry, conservation biology, soil science, physical geography, applied climatology or public administration, are more overtly political questions. To the extent that they and others are turning this idea into a writ of empowerment to preside over the declaration, and then implementation of, an ecological state of emergency, it matters (Bierman, 2007, pages 325–37). 'Letting go' of visions of ecological catastrophism as a natural accident is crucial because the world is beyond the natural and the accidental. In actuality, it has many technified dimensions in which 'the deciders', who specialise in adaptations for climate change, always try to 'right-size' carbon-intensity, growth prospects and globality for the few, while geoengineering schemes develop eco-managerialist designs to decarbonise, degrow and deglobalise everyday life for the many (Westley et al., 2011, pages 762–80).

The growing coalition of specific intellectuals in resilience science closely parallels Foucault's (1980, page 95) appraisals of productive power: '[T]here are manifold relations of power which permeate, characterize, and constitute the social body, and these relations of power cannot themselves be established, consolidated nor implemented without the production, accumulation, circulation, and functioning of a discourse.'

This 'economy of discourses of truth' as it is now operating as the basis of many anthropocenic applications for earth system science appears to provide the authoritative alibi for a directorate of planetarian management to assume command and control by generating policy-ready findings for stronger waves of allegedly sustainable development (Luke, 2005, pages 228–38). As Ehlers and Krafft maintain, the resilience challenge is immense, important and imperative. While it will be global in scope and authority, 'Earth System Science has to provide place-based information by analyzing global and regional processes of Global Change and by translating the research findings into policy-relevant results' (2010, page 10).

Resilience science, therefore, plays 'on the catastrophic to criticize a more patient science for its adherence to the most "likely" scenario and its failure to give us the truth of extreme climate changes', which in turn generates added affirmative momentum 'for suspending the law' to police and protect resilience (Marzec, 2012, page 43). The reduction of global ecologies in environmentality practices to planetary infrastructure also tacitly legitimates the current need for 'building capacity' in all social-natural systems to 'absorb disturbance' as well as 'maintain function' (Walker and Salt, 2012, page 1) for the high-value populations and places favoured in resilience science operations. The restoration of the Holocene will not occur, but the engines of restorative invention must power up enough to maintain those illusions in the Anthropocene at least for a few as everyone suffers along the fronts of many disturbances.

Centring policy discourse and practice on fundamental changes in the energy economy of human culture now sits at the heart of resilience science. As the energy, information and material capacity of ultra-modernising 'Society' mobilises multiple sources of power to rebalance 'the biological' against 'the historical' in humanity's engagement with 'Nature', the brittle stability of recent history shatters. It is no longer a stable baseline for the resilience machine coming to constitute what Foucault would imagine as a 'new historical *a priori*'. Resilience, as Holling shows, is never wholly described, because it is in many ways a system of 'sustainable degradation' (Luke, 2006, pages 99–112). Indeed, its incompleteness – at what could be the Holocene to Anthropocene turn – makes it far more possible to initiate and sustain 'a series of complex operations that introduce the possibility of a constant order into the totality of representations. It constitutes a whole domain of empiricity as at the same time *describable* and *orderable*' (Foucault, 1970, page 158).

Now, as Foucault might assert, the 'framework of thought' behind the restorative targets of regaining Holocene against the radical disruptions of the Great Acceleration in the Anthropocene seeks the key boundary conditions. In other words, it 'delimits in the totality of experience a field of knowledge, defines the mode of being of the objects that appear in that field, provides man's everyday perception with theoretical powers, and defines the conditions in which he can sustain a discourse about things recognized to be true' (Foucault, 1970, page 158), and resilience thereby gains even greater depth, breadth and colour.

The mixed and mangled histories of the Earth's natural and social forces make any full restoration a very problematic task. What equals that which can, or should, be restored amidst overfished oceans, heavily polluted skies, overburdened farm lands, destroyed animal species, corporate hybridised plants, industrially timbered forests as the summation of 'Nature' restored for 'Humanity'? As Roberts argues with regard to the short 11 to 12 millennia of the Holocene, 'in reality many ecosystems are far from being wholly "natural", and instead owe their distinctive character to particular manners of land use or other human actions ... [F]or most ecosystems it is therefore effectively impossible to study environmental history separate from cultural history, and vice versa' (Roberts, 1998, page 251). In this respect, how the restoration engines spins within resilience science is not much more than another organised campaign to discipline what 'the deepest rupture in the history of the environment', namely, 'the failed Americanization of the world' (Radkau, 2008, page 250).

Deformed Modernity and Resilient Subjects

Perhaps resilient subjects are not ordinary neoliberal agents, but rather another series of subjects in deformed modernity that have received too little attention. Liberal democratic capitalism clings to its founding visions, which are ironically very dark. Hobbes' vision of the Leviathan as an apparatus to escape violent struggles in the state of nature, in which life is 'solitary, poor, nasty, brutish, and short' (Hobbes, 1994, page 76), only masks the emergence of states whose constitution does little more than aggressively organise how solitary, poor, nasty, brutish and short life becomes in the gaze of its sovereign authority. And Locke's vision of civil society in need of government to protect the capacity of citizens to enjoy more 'life, liberty, and the pursuit of property' is little more than a biopolitical bandage over this permanent precarity (Locke, 1980, pages 33–46). With a global economy running on colonialism, slavery and dispossession along with a new state system typified by imperial aspirations, continuous war and tyrannical whims, Locke's social contract gives its parties the rights to coerce each other and rebel against those infringing their freedoms along with a warrant of industry to destroy, dominate and/or dispossess those not using their labour most effectively amidst Nature's bounty to create more property. Likewise, Rousseau's sense of civil society acknowledges how enforcing property rights over people and territory is indeed 'the origin of inequality' at the core of human degradation, nature's despoilation and state domination (Rousseau, 2012, pages 27–120). Once put into position, however, the best outcome anyone can attain is complete and full participation as subject and citizen in the moving matrices of the general will, 'forced to be free' (Rousseau, 2012, pages 219–36) while he or she works to avoid extirpation.

Bauman appears to ignore these rough edges of modernity in which Marx and Engels saw 'all that is solid melts into air' (Tucker, 1978, page 476). Forgetting Hobbes and Locke, Bauman (2007, page 26) writes:

> Fear is arguably the most sinister of demons nesting in the open societies of our times. But it is the insecurity of the present and the uncertainty about the future that hatch and breed the most awesome and least bearable of our fears. That insecurity and that uncertainty, in their turn, are born of a sense of impotence; we no longer seem to be in control, whether singly or collectively.

This sort of meditation reeks of modernist chronocentrism and Whiggish utopianism. When was fear not a feature of any society – open or closed – in modern times? Modernity's impositions of its instrumentally rational order upon all substantially rational non-/pre-/un-orders rarely, if ever, create a secure present, a certain future or a bearable fear.

As Rousseau opines in his first two discourses, once humans are torn from Nature, individual and collective fear, insecurity, powerlessness and/or uncertainty is the permanent product of 'progress in the Arts and Sciences', culminating in permanent inequality, pervasive impotence and precarious incapacitation at the hands of this social order's endemic industries of fear, mystification and want (Rousseau, 2012). Resilient individuals and collectives actually have had to forget freedom for centuries rather than decades as they adapt to the catastrophes of commercialisation, genocide, industrialisation, empire and urbanisation. Psychological resilience in times of modernisation has been seen as the capacity to overcome 'future shock' (Toffler, 1971) as well as the social recognition that individual freedom grows jaggedly from the recognition that 'society must be defended' (Foucault, 2003). The conditions of human life are rarely ever all good or bad, but they are always dangerous. Nevertheless, the restoration engine at the core of resilience spins up behind these embedded existential expectations of fear, injury, loss and risk being redressed sometime, somewhere, somehow by a return to full normality.

At the same time, Allenby's perspectives on the planet's history completely question the ideal conditions that resilience thinking would target for restoration.

> The Earth has become an anthropogenic planet. The dynamics of most natural systems – biological, chemical, and physical – are increasingly affected by the activities of one species, ours ... Although this process has been accelerated by the Industrial Revolution, 'natural' and human systems at all scales have in fact been affecting each other, and coevolving, for millennia, and they are more coupled than ever. Copper production during the Sung dynasty as well as in Athens and the Roman Republic and Empire are reflected in deposition levels in Greenland ice; and lead production in ancient Athens, Rome, and medieval Europe is reflected in increases in lead concentrations in the sediments of Swedish lakes. The build up of carbon dioxide in the atmosphere began not with the post-World War II growth in the consumption of fossil fuel, but the growth of agriculture in, and thus deforestation of, Europe, Africa, and Asia over the past millennia. Humanity's impacts on biota, both directly through predation and indirectly through the introduction of new species to indigenous habitats, have been going on for centuries as well.
>
> *(Allenby, 2013, pages 9–10)*

This perspective is provocative in as much it underscores how the Holocene has been drifting on tides of anthropogenic ecological degradation for millennia. Moreover, the anthropocentric reconstruction of humanity's work since the rise of civilisation is, to a considerable extent, a crypto-resilience in action. Its 'transition' planning narrative works to solidify other capacities, forestalls different disturbances and sustains social-ecological functions in unanticipated ways on an unintended scale (Barry, 2012). Similarly, it exposes both the flow and force of resilience recoil for those aspiring to run the restoration engines.

Restoration Impossible

Restoration itself will never be flawless, identical or total. Restoring anything will always miss something, replace entire parts or pieces with newer elements and fail to recover in toto what held true prior to the disturbances triggering a restorative intervention. Restoration biology, historic restoration, restoration botany and artistic restoration all sound so plausible, as if the intended end state can be attained. They all fail, because the project cannot succeed.

At best, then, the restoration engine is a biotic and climatic wager that pans out by creating biotic and atmospheric currency exchanges instead of actually returning the same species and spaces lost at any given locale to some original biospheric balance. Hence, it must give 'attention to "process" rather than final outcomes' since rewilding processes can only at best re-establish and enhance '"ecology's dynamic interactions" in all their variety' (Davies, 2016, page 201). The restoration engine, then, only runs routines, produces outcomes or generates ecological interactions, but the bounce-back has new springs, different tension, more play and less fidelity. Yet, as an *ingenium*, as Latin discloses, the engine 'begets', 'procreates' or generates enough to excite whole movements of restorationists. Whether it is bringing hybrid bison/cattle 'beefalo' back to abandoned grain farms on the Great Plains or Asian chestnut varietals to the niches American chestnuts occupied in Appalachia, these engineered and bounded micro-genesis events get trucked, bartered, exchanged and traded in the resilience markets as threshold-event deflectors, containers or aggregators (Steffen et al., 2011, pages 739–61). The hidden recoil in resilience cannot be escaped. Biopolitics becomes cyberpolitics, working not to preserve all biota as such, but rather to estimate, and then simulate, the basic enviro-functionalities those now lost biota once performed with neobiotic capacitors to enable disturbance management (O'Connor, 2015; Wray, 2017).

Recovery is neither getting back what was lost nor regaining better bits of an original condition. Instead it is the adaptive actions of reclaiming enough of a replacement, supplement or equivalent to make up for the lost, gaining better or even good functionality degraded or destroyed. The restoration engine is only a ruse of recovery that accepts compensatory, alternate and varied new gains to take back bits of systemic integrity and stability in exchange. Restoration seems convincing in as much as 'what was' finds its simulacra of reconstruction; but, like the post-1945 restorations of Berlin, Coventry, Dresden, Kaliningrad, Nagasaki or Warsaw, all reconstructions are only close replacements, never the exact resurrection of an original appearance, feel or presence. Recuperative action is activity that takes (*capere*) back enough to meet resilience-science criteria, making communities or landscapes adaptive and functional.

Such green governmentality interventions rest necessarily on shifting streams of 'supplements'. In fact, all disturbances degrade the original condition, and their functionality can be slowed to a halt. The ruse of recovery in resilience, as pressed through restoration engines, is the supplementality that lessens the gaps. Indeed, supplements make up for a deficiency and enable the fresh acts that constitute coping with loss rather than copying what was lost. All of these substitutions, switches or shifts in supplementality demand tremendous craft; here is where and how the power/knowledge of resilience science congeal today in artful anthropocenarian garb.

Living in disruptive times and degraded spaces, therefore, is not necessarily always dire. Hunter (2007, pages 316–18) suggests:

> Many animal and plant species have adapted to the new stresses, food sources, predators, and threats in urban and suburban environments, where they thrive in close proximity to humans. Their success provides researchers with valuable – and sometimes unexpected insights – into evolutionary and selective processes. Because these adaptations have had to be rapid, cities are, in some respect, ideal laboratories for studying natural selection.

Such contradictory conclusions say much about living with greater resilience. It is a lot like living close to the lost conditions of the Holocene, only now its many human inhabitants are capable, on the one hand, of accepting contemporary environmental depredations while, on the other hand, adapting to the biodiversity loss of the Great Acceleration. Regrettably, the fixations on rapid climate change push far too many scientists, policy experts and decision makers toward tussling over how to grant themselves command and control during this ecological state of emergency.

Crutzen (2010, page 17) is clear about his agenda: 'Exciting but also difficult and daunting tasks lie ahead of the global and engineering community to guide mankind toward global, sustainable, environmental management into the Anthropocene'. Hoping to construct transition towns or resilience redoubts, these social forces ignore how it is far too late to restore the planet to conditions that prevailed in 1990, 1970, 1950 or 1900. Even if that return were possible, it would not lessen, mitigate or end the impact of the Great Acceleration. On this point, Hunter (2007, page 216) affords a constructive, useful, if unintended, assessment: 'Urbanization does not preclude the development of teeming habitats; rather than being confined to remote areas and wildlife parks, they can be found in densely populated areas … Indeed, business parks exemplify how an artificial environment can be exploited and enhanced by conservation initiatives.' Hunter and others are thrilled about the repopulation in business parks, cities and suburbs with new biota to be wildlife preserves, but they miss the main point of these 'working landscapes' (Cannavò, 2007).

The discourses of conservation biology, resilience science, sustainability studies or complex coupled systems management only entrench the deepest manifestations of their knowledge as power, or power calling forth their knowledge, to further exploit the natural and social zones of the Earth in the evolving power grids of environmentality that require happier and healthier ensembles of action for the planetary 'business park'. Hence, 'if we are to comprehend, let alone move toward grappling with, the world we are continually remaking, we must get behind the idea that we are imposing our intent, our purpose, on the future. Religion may be the opiate of the masses, but "cause and effect" is the opiate of the rational elite' (Allenby and Sarewitz, 2011, pages 70–1).

Despite the hope he places in 'degrowth', Latouche agrees that today's planetary urbanisation is 'a sort of Megamachine that has now become anonymous, deterritorialized and uprooted from its historical and geographical origins' (Latouche, 1996, page xii). Resilience in the Anthropocene also can articulate itself with varying arrays of keeping or restoring urban cultural alternatives, effectuating a 'worldwide standardization of lifestyles … which is imposing a one-dimensional, conformist way of living and behaving on the ruins of abandoned cultures' (Latouche, 1996, page 59). Absorbing recoil is clearly the benchmark of resilience conformity, since gaining this capacity is where everyone hopes things will kick back, if recovery begins.

Resilience as Performative Recoil Response

To conclude, resilience, as the output of restoration engines is actually defined in innovative terms, such as shorter-term responses to abrupt shocks in natural disasters or weather emergencies and the longer-term adaptive capacity of individuals to cope with unsteady economic conditions or recurrent market disequilibria. In all these circumstances, however, the basic principles resting within the core of resilience, such as recovery, recuperation or restoration, allow 'bounce-backability' always to be modulated in its reactive or proactive range of recuperative interactions. It rhetorically aims to recapture former states of relative balance, normality or steadiness, but it operationally accepts extreme imbalance or abnormality alongside enhanced shakiness. Recoil again is the norm, but it is receiving the repetitive hard kick of change, which numbs most people against fixating on the implausibility of getting through tough times and places to that which prevailed before disasters hit.

Sophisticated versions of resilience science admit the world is contingent, complicated and chaotic; hence, turbulence will at best attain another relatively ordered state that could enable a stabilised adaptive reset, emulate what was lost with good supplements or approximate the best supplements for the destroyed. Less sophisticated variants of resilience science might see the environment as still predictable, placid or permanent. Believing that the goal for resilient responses is recapturing a more normal condition specified in terms of the ability to bounce back into existing prior patterns is convenient but mystifying.

Strangely enough, then, resilience thinking fits well in the logics of performativity after the Nature of the Holocene goes away for good. Human beings are now the key anthropocenarian species. Their labour is creating or cultivating a diverse array of anthropoceneries with tremendous terraformative impact on their immediate surroundings, the planet at large and now deep geological time. The net outcome is destabilising, desecuritising and degrading as this radical endangerment defines the everyday existence of all terrestrial life forms. Whether one sees humans adjusting to orderly turbulence or just predictable patterns in moments of abrupt shock to constant stabilities, the agendas of performativity remain in play.

Amid these chaotic conditions, people exert imperatives to maintain productivity, safety and stability in bouncing back to patterned order or coping constantly with chaotic variations. Most ironically, resilience preparedness for worlds of insecurity and instability is sensitising individuals and groups to believe disturbance avoidance and functional maintenance for strong bounce-back is possible. Yet any recuperative bounce-back only returns them to new panarchic conditions of insecurity and instability (Gunderson and Holling, 2002). Rebuilding a town in the tornado belt of the American Midwest is a recuperative restoration of human settlement; yet any restored town is still in the tornado belt, and that zone's storminess is increasing. Here is the raw deal in common recoil.

Resiliency is another side of green governmentality at work producing particular subjects with the knowledge and power needed to maintain what was, while pushing up, on, out to attain fresh productive advances in sustaining the development of ever more degraded material worlds needing greater governance.

The jury, one could say, is still out. Even so, many putative ecological practices, which are produced every day in sustainability offices, studies and policies around the world, are another instance of neoliberalism's efforts to mobilise ideas such as popular government, individual responsibility and free markets to serve other hidden state and corporate agendas, such as adapting to the yet-to-stabilise Anthropocene. Within the vaporous valorisation of being/becoming sustainable players in a resilient society, the governmentalising principle of individual responsibilisation has been put into action. The opportunities for managing the conduct of our conduct reveal not what is promised as freedom but what is being imposed as necessity. Governmentalising people, responsibilising the individual and marketising freedom in the panarchy of the Anthropocene puts economies and societies under closer and tighter planetarian accountancy to anchor risk and enhance resilience (Luke, 2009, pages 129–59). Here governance mediates itself in materiality itself, the embedded directives and layered legislation of which steer allegedly free choice toward responsibilising individuals 'to save the Earth'. When these resilient subjects freely choose which markets to enter, the objects around which their subjectivity is shaped thereby governmentalise them.

The restoration engine humming at the heart of the resilience machine exemplifies these contradictions. Every risk-management strategy to enhance resilience implicitly concedes the restoration promised in such strategies is mythic, which compounds all changes multiple times into other complex managed risks. One must, in turn, cultivate continuously the capacities of disturbance description, functional maintenance and practical resilience needed to bounce back from its known unknown hazards because bigger and badder risks lie ahead in the Anthropocene. Ordinary encounters with lesser threats essentially cannot steel one to endure those ever worsening ordeals of planetary degradation.

In this respect, 'learning from Louisiana' is helpful. From the times of the Great Floods in the Mississippi basin in 1927 to those experienced state-wide there during August 2016, real restoration never can be gained from anthropogenic settings (Grant, 2015; Barry, 1998). At best, one judges each disaster by how high the waters rose, where the mud stopped or which friends and families were lost because these hard kicks recoil in pounding back groups and individuals with every new cycle of resilience practice. Their constant hidden recoil frames resiliency as the uneasy coexistence with constant hazard, raw risk and dark destruction whose restoration engines never steer anyone to safety, risklessness or security. Instead, they pull everyone toward adapting to even greater hazards, rougher risks and deepening destruction.

References

Allenby, B. (2013). *Reconstructing the Earth: technology and environment in the age of humans.* Washington, DC: Island Press.

Allenby, B. and Sarewitz, D. (2011). *The techno-human condition.* Cambridge, MA: MIT Press.

Barry, J.M. (1998). *The Great Mississippi Flood and how it changed America.* New York: Touchstone.

Barry, J.M. (2012). *The politics of actually existing sustainability: human flourishing in a climate-changed, carbon constrained world.* Oxford: Oxford University Press.

Bauman, Z. (2007). *Liquid times: living in an age of uncertainty.* Cambridge: Polity Press.

Bierman, F. (2007). 'Earth system governance as a cross-cutting theme of global change research'. *Global Environmental Change*, 17: 325–337.

Biro, A. (2015). 'Good life in the green house? Autonomy, democracy, and citizenship in a warmer world'. *Telos*, 172(Fall): 15–37.

Brenner, N. and Schmid, C. (2011). 'Planetary urbanization', in M. Gandy (ed.), *Urban Constellations*. Berlin: Jovis, pp. 10–13.

Cannavò, P. (2007). *The working landscape: founding, preservation and the politics of place.* Cambridge, MA: MIT Press.

Crutzen, P.J. (2010). 'The "Anthropocene"', in E. Ehlers and T. Krafft (eds), *Earth system science in the Anthropocene: emerging issues and problems.* Berlin: Springer Verlag, pp. 13–18.

Crutzen, P.J. and Stoermer, E.F. (2000). 'Have we entered the "Anthropocene"?'. *IGBP-Global Change*, 31 October.

Davies, J. (2016). *The birth of the Anthropocene.* Berkeley, CA: University of California Press.

Davoudi, S. (2014). 'Climate change, securitisation of nature, and resilient urbanism'. *Environment and Planning C: Government and Policy*, 32(2): 360–375.

Davoudi, S. (2016). 'Resilience and governmentality of unknowns', in M. Bevir (ed.), *Governmentality after neoliberalism.* New York: Routledge, pp. 152–171.

Ehlers, E. and Krafft, T. (2010). *Earth system science in the Anthropocene: emerging issues and problems.* Berlin: Springer Verlag.

Evans, B. and Reid, J. (2014). *Resilient life: the art of living dangerously.* Cambridge: Polity Press.

Foucault, M. (1970). *The order of things: an archaeology of the human sciences.* New York: Vintage.

Foucault, M. (1980). *Power/knowledge: selected interviews and other writings, 1972–1977.* New York: Pantheon.

Foucault, M. (2003). *Society must be defended.* New York: Picador.

Foucault, M. (2007). *The politics of truth.* New York: Semiotext(e).

Gillis, J. (2013). 'Study of Ice Age bolsters carbon and warming link'. *New York Times*, 1 March: A4.

Grant, R. (2015). *Dispatches from Pluto: lost and found in the Mississippi River Delta.* New York: Simon and Schuster.

Gunderson, L.H. and Holling, C.S. (2002). *Panarchy: understanding transformation in human and natural systems.* Washington, DC: Island Press.

Hobbes, T. (1994). *Leviathan: with selected variants from the Latin edition of 1668*, ed. E. Curley. Indianapolis, IN: Hackett Publishing Company.

Holling, C.S. (ed.) (1978). *Adaptive environment assessment and management.* London: Wiley.

Hunter, P. (2007). 'Analysis: the human impact on biological diversity'. *European Molecular Biology Organization*, 8(4): 316–318.

Kolbert, E. (2013). 'Enter the Anthropocene: age of man', in E. Ellsworth and J. Kruse (eds), *Making the geologic now: responses to material conditions of contemporary life.* Brooklyn, NY: Punctum Books, pp. 28–32.

Latouche, S. (1996). *The Westernization of the world*. Oxford: Blackwell.
Locke, J. (1980). *Second treatise of government*, ed. C.B. Macpherson. Indianapolis, IN: Hackett Publishing Company, pp. 33–46.
Luke, T.W. (2005). 'Neither sustainable nor developmental: reconsidering sustainability in development'. *Sustainable Development*, 13(4): 228–238.
Luke, T.W. (2006). 'The system of sustainable degradation'. *Capitalism Nature Socialism* 17(1): 99–112.
Luke, T.W. (2009). 'Developing planetarian accountancy: fabricating Nature as stock service, and system for green governmentality'. *Current Perspectives in Social Theory*, 26: 129–159.
Luke, T.W. (2015). 'On the politics of the Anthropocene'. *Telos*, 172(Fall): 139–162.
Marsh, G.P. (1965 [1864]). *Man and nature*. Cambridge, MA: Harvard University Press.
Marzec, R.P. (2012). *Militarizing the environment: climate and the security state*. Minneapolis, MN: University of Minnesota Press.
McNeill, J.R. and Engelke, P. (2014). *The great acceleration: an environmental history of the Anthropocene since 1945*. Cambridge, MA: Harvard University Press.
O'Connor, M.R. (2015). *Resurrection science: conservation, de-extinction and the precarious future of wild animals*. New York: St. Martin's Press.
Radkau, J. (2008). *Nature and power: a global history of the environment*. New York: Cambridge University Press/German Historical Institute.
Resilience Alliance (2010). *Assessing resilience in social-ecological systems: workbooks for practitioners*, Version 2.0. Available at: www.resalliance.org/3871/php
Roberts, N. (1998). *The Holocene: an environmental history* (2nd edn). Oxford: Blackwell.
Rousseau, J.-J. (2012). *Rousseau: the basic political writings: discourse on the sciences and the arts, discourse on the origin of inequality, discourse on political economy, on the social contract, the state of war*, ed. and trans. D.A. Cress (2nd edn). Indianapolis, IN: Hackett Publishing Company, pp. 27–120.
Ruddiman, W.F. (2005). *Plows, plagues, and petroleum: how humans took control of climate*. Princeton, NJ: Princeton University Press.
Steffen, W., Persson, A.Deutsch, L. et al. (2011). 'The Anthropocene: from global change to planetary stewardship'. *Ambio*, 40(8): 739–761.
Toffler, A. (1971). *Future shock*. New York: Bantam Books.
Tucker, R. (1978). *The Marx-Engels Reader* (2nd edn). New York: Norton.
Walker, B. and Salt, D. (2006). *Resilience thinking*. Washington, DC: Island Press.
Walker, B. and Salt, D. (2012). *Resilience practice: building capacity to absorb disturbance and maintain function*. Washington, DC: Island Press.
Westley, F., Olsson, P., Folke, C. et al. (2011). 'Tipping toward sustainability: emerging pathways of transformation'. *Ambio* 40(7): 762–780.
Wray, B. (2017). *Rise of the necrofauna: the science, ethics, and risks of de-extinction*. Vancouver, BC: Greystone Books.
Žižek, S. (2000). *The fragile absolute: or, why is the Christian legacy worth fighting for?* London: Verso.

INDEX

Note: italic page numbers indicate figures; numbers preceded by *n* refer to chapter endnotes.

accountability 7, 132, 133, 134
actor network theory 18
adaptation 166–170; participatory 110, 111, 112, 113, 119, 120–121
adaptation machines 4, 110–121, 150; and affective relations 111, 113–114, 116, 117–118, 119, 121; as assemblages 113, 118–119; and biopolitics 114, 115, 117, 119; catastrophe insurance as *see* CCRIF; and diagrammatic interventions 111, 113, 114; and environmental governance 112–113, 114, 117; and ethico-aesthetic paradigms 4, 111, 119, 121; and fishing community initiative 120–121; future radical possibilities for 118–121; as self-organising systems 113
adaptive capacity 111–114, 118–119, 145
adaptive cycle model 22, 150, 151, 154, 155
adaptive management 112, 155
adaptive resilience 10
Adey, Peter 94, 95, 97, 118
Adger, Neil 104
AE (assisted evolution) 54–55, 57
aesthetics of community 97, 98, 102
aesthetics of resilience 95, 97, 98, 118
affective relations 4, 68–69, 95; *see also under* adaptation machines
African Americans 23, 68
Agamben, G. 104

agency 2, 32, 39, 110, 113, 154; and community resilience 64; and smart bodies 46, 47
Al-Qaeda 31, 32
Alexander, D.E. 20
alienation 186
Allenby, B. 202, 204
Anderson, B. 96
Anderson, Benedict 102
Anguelovski, I. 167
Anthropocene 9–10, 56, 104, 179, 198–200; and 'Great Acceleration' 192, 198, 200, 204; and neoliberalism 191–192, 194, 195–196, 206; and resilience 194–196
anti-microbial resistance 57
Aradau, C. 84
Aranda, K. 2–3
Arendt, Hannah 5, 177, 178, 181, 182–183, 186, 189
ARPANET 48
Åsberg, C. 2
assemblage diagram 24
assemblages 1, 2, 3, 6, 7, 9, 18–20, 145, 154, 155; adaptation machines as 113, 118–119; collective 20; and community resilience 63, 67; and complexity 20; definition/uses of term 18–19; and emergence/exteriority 19, 21; ethos of 118–119; fuzzy boundaries of 2, 19;

indeterminate/complex interactions of parts in 19–20; machinic *see* machinic assemblages; and panarchy 22; and resilience machine 20; and resistance *see* resisting resilience machines; risk 115; and smart bodies 45–46; and vulnerability 11
assisted evolution (AE) 54–55, 57
Athanasiou, Athena 30
Augustine 186
austerity 17, 179
auto-poetic nature 44, 46, 47, 51, 55, 56
automated resilience 7
autonomisation 63

Bacon, Francis 14, 20
Bakan, J. 70, 71, 74
Barrios, R.E. 70
Barry, J. 181
Basque Country (Spain) 73
Bauman, Zygmunt 6, 201
Bautista, Eddie 171
Beck, U. 148, 180, 187
Beilen, R. 152
Belfast (UK) 25
Berkes, F. 21
Bertelsen, L. 121
Bier, Ethan 53
big data 7, 156n2
bioengineering 50–59; cellular manipulation 51–52; and coral reefs 54–55; CRISPR/gene drive technologies 52–53, 56, 57, 58; dangers of 57–58; and diabetes treatments 51; fly genetics 53–54; and pre-emptive thinking 56–57; tissue engineering 50–51, *see also* smart bodies
biology 15, 22, 52; new 47; and security 43
biomedicine *see* bioengineering
bioperfection imperative 39, 41
biopolitics 114, 115, 117, 119, 129–130, 193, 196, 203
biotechnologies 52; and biomimicry 44–45; and governance 46, *see also* smart bodies
Biro, A. 193
Blake, William 182
Bloomberg, Michael 160, 160–161, 164, 170, 171
Bobbio, Norberto 23
Boelens, L. 146
Boschma, Ron 147
Botterill, L.C. 96
bottom-up initiatives 8, 84, 87–89, 90, 133, 135, 138, 141, 153

bounce-back-ability 5, 6, 23, 85, 97, 149, 195; and hidden recoil of resilience 193–194, 202, 203, 204, 205–206
boundaries 22; and community 99; fuzzy 2, 19; and imagination 34, 37; and resilience thinking 10, 96, 97–98
Bradley, Jennifer 164
Brandenberg (Germany) 22
Brave New City 87
Brazil 73–74
Breaking Bad (TV series) 96
Brecht, Bertold 185
Brenner, N. 198
Brent, Jeremy 100
Bretherton, Luke 178, 184–185, 186–187, 188
Brickell, K. 101–102
Bristow, G. 154
Bronx (New York) 161–162, 163, 170, 171
Brooks, David 68
Brown, L. 9
Brugmann, Jeb 178
Bud, R. 44–45
Bush, George W. 31
Butler, Judith 30

C3 (command/control/communications) 47–48, 85
California, San Diego, University of 53–54
California San Francisco, University of 51
Callon, Michel 18
Campanella, R. 69
Canterbury Earthquake Recovery Authority 85
Canterbury (New Zealand) *see* Christchurch, disaster recovery in
capacity building 110, 133, 139, 192, 194, 199
capital accumulation 5, 116, 117, 153
capitalism 38, 65, 70–71, 81, 83, 186; late 119, 180; post- *see* post-capitalism; resilience of 153; terminal/'end times' 188
Capra, F. 14, 15
Caribbean 4; fishing community initiative in 120–121, *see also* Jamaica
Caribbean Catastrophic Risk Insurance Facility *see* CCRIF
Castree, Noel 102
catastrophe 33, 57, 180–182, 192–195, 197; biological 45, 55
catastrophe insurance *see* CCRIF
catastrophe modelling 116
catastrophe risk *see* disaster risks

catastrophism 177, 180–182, 192–194, 199
CCRIF (Caribbean Catastrophic Risk Insurance Facility) 4, 114–117; and biopolitics 115, 117; and catastrophe modelling 116; and global financial markets 115, 116, 117; and parametric insurance 115–116, 121n1, 122n2; and risk price/value 116
Central City Recovery Plan 88
Chakrabarty, D. 104
Chandler, David 49–50, 96, 97, 105, 107, 148, 155
change 4, 64, 126–127, 168, 195; global 177, 199
Cheney, G. 72
Christchurch, disaster recovery in 4, 81, 84–89; and central government 85; and community resilience team 85; and community-led/transformational projects 84, 87–89, 90; extent of disaster 84; and local government 84, 85, 89–90; and neoliberalism 85; and self-sustaining/coping qualities 85–86
Christianity 183, 184, 185, 186
Church, George 55
cities 2, 7; evolution of 15; good 5, 177–178, 182, 184; 'good enough' 178, 184–187; and planning *see* urban planning; and political machines 16; relational aspects of 5; resilient *see* resilient cities; urban resilient machine; resource flows in 6; walking/moving in 7, *see also* urban development; urban growth machine; urban politics
citizens/citizenship 4, 18, 85, 102, 127, 132; and self-determination 129, 138, 140
city mayors 5, 69, 85, 159, 160
city resilience indices 7
civil society 5, 127, 135–136, 141, 200–201
Clements, Frederick 99
Clifford, T. 16
climate change 2, 13, 62, 81, 179, 193, 194–195, 197–198, 204; adaptation *see* adaptation machines; and imagination 33; insurance mechanism for *see* CCRIF
co-evolution 49, 88, 181
co-option 8, 18, 72, 82, 89, 165
coastal areas 160–161, 162, 163, 166, 197, *see also* waterfront land
Cohen, D. 56
Cohen, M.D. 147
Cold War 191, 194

collective assemblages 20, 24, 25
Collier, S. 115
command/control/communications (C3) 47–48, 85
communitarianism 24, 73, 186
communities 2, 6; appropriation by resilience machine of 24; and governance 66, 67; and loss/decline 64–66; and openness/change 4; and planning 5; and social insurance/risk 23, 66–67; translocal 102; versus society 3, 65, 66; virtual 97, 103–104, 107
community 94–107; aesthetics of 97, 98, 102; and boundaries 99; as co-becoming 105–106, 107; consensus notions of 98–99; critiques of 96, 97, 99–106; desirability/values of 99–100; and environmental crisis 104–105; as global condition 104; imagined 102–103; and justice 100–101; and place identities 99, 101–102; and resilience diagram *see* resilience diagram; and resilience thinking 94–98, 105, 106–107; virtual 97, 103–104, 107
community development 66, 100, 130, 160, 171–172
community identities 100–101, 102, 104
community organisations 170–172
community relations 6, 18, 62, 75, 166
community resilience 4, 62–76, 106–107; and alternatives to neoliberalism 71–74, 75–76; and 'community-based' problems 62; and corporations 70–71, 72; defined/origin of term 64; and equity/social capital 64, 65, 69; interrelated factors in 64; and liminality 73, 74, 75; and neoliberalism 3–4, 63, 67, 70, 75–76; and RAND toolkit 69, 71; and resilience thinking 95–96; schematics of *see* resilience diagram; and self-reliance/-responsiveness 23–24, 64, 82; and the 'social' 22, 66, 74–75; and solidarity *see* solidarity; and worker cooperatives 72–74; and worker-led takeovers 73–74, *see also* New Orleans, disaster recovery in
competition 47, 70, 74, 144
complex systems 1, 5, 8, 20, 21, 95, 138, 148, 149; Panarchy model of 21–22; self-organised 96, 150, 153; and smart bodies 52; systems of 196–200
complexity 2, 5, 8, 20, 21, 22, 24, 168; and assemblages 20; in biology/physics 15; and emergence 147–148; governing

49–50, 105, 144, 145, 147, 154; and networks 48; and smart bodies 46; and urban planning 144–147, 152
computers 38, 47, 48, 52, 168
Connolly, W. 25
control 8, 10, 37
Copernicus, Nicolas 14
coral reefs 54–55
Corburn, Jason 171
corporate social responsibility 73
corporations 3, 6, 7, 59, 70–71, 74; and Resilience Alliance 13, 21; and solidarity 72–73
Cote, M. 168
counter machine 25
CPI (Community Parks Initiative, New York) 170
Crisanti, Andrea 56
crisis politics *see* post-disaster recovery
CRISPR (Clustered Regularly Interspaced Short Palindromic Repeats) 52–53
Crist, E. 10
critical political geography 150–154; and institutional dimensions/power relations 151–152; and justice/democracy 154; and neoliberalism 153–154; and scalar/relational approach 151
critical theory/scholarship 32, 33, 41*n*2, 75, 182
Crutzen, Paul 195, 204
cummunitas 73–74
currency, alternative 88, 90*n*2
cybernetics/cyborgs 2, 46, 47, 48, 54, 96, 103
Cyrulnik, Boris 31, 32

Dalziel, Lianne 85
DARPA (Defense Advanced Research Projects Agency) 58
Darwin, Charles 15
Datta, A. 101–102
Davis, M. 180, 181–182
Davoudi, S. 21, 22, 24, 97, 145, 150, 177
de Blasio, Bill 5, 161, 164, 169–170, 173*n*5
Dean, Mitchell 115
Debrix, François 7
DeLanda, Manuel 19, 20, 24
Delanty, G. 99, 103
Deleuze, Gilles 18, 19, 20, 66, 113, 117; and typology of images 39–41
Delsarte, Francoise 36
democracy 5, 13, 15, 82, 154, 179; encroachments on 82, 85, 89, 90*n*1

depoliticisation 82, 83, 86, 89, 110; and disaster management 112; of environmental governance 4, 112–113, 118
Derickson, K.D. 153
Descartes, Rene 14, 15
Devadas, V. 104
development projects 1, 110
diabetes 51
diagrammatic interventions 95, 111, 113, 114, *see also* resilience diagram
Dill, K. 82
disaster management 7, 13, 24, 111–112, 156*n*3; and catastrophe insurance 117; and resilient cities 127, 130, 131, 140
disaster preparedness 62, 164
disaster recovery *see* post-disaster recovery
disaster risk management 13, 24, 131
disaster victims 24–25
disasters, future/potential 31–32, *see also* catastrophism; climate change
disasters and transformation 68, 71, 81, 83–84
disease 54, 56, 57
Dolan, F.M. 183
Donzelot, J. 67
Downes, B.J. 23
Drosophilia Melanogaster 53
du Plessis, Gitte 57
Durkheim, Emile 65

Eagleton, T. 5, 178, 184, 188
earthquakes 20, 115
ecology 21, 22, 99
economic growth 8, 16–17, 115, 128, 137, 161, 191–192; diversification of 17
ecosystems 6, 10, 13, 16, 80, 131, 151, 194, 200; and assisted evolution 54–55, 57
education 23, 68, 73, 164; Freierian pedagogy 111–112
Ehlers, E. 199
Eisenstein, M. 51, 52
elites 16, 17, 188
employment 72–74; and liminality 73, 74, 75; security 73; and worker cooperatives 72–73; and worker-led takeovers 73–74
empowerment 64, 68, 74, 100; and modernist political imaginary 111, 112; of women/girls 129
Engels, Friedrich 185, 201
entrepreneurialism 17, 23, 70, 71, 74
entropy 15, 181, 194
environmental crisis 13, 104–105, 177–178, 188, 191, 194–195

environmental governance 4, 112–113, 117
environmental justice 5, 160, 161–162, 164–166, 171–172, 173
Environmental Justice Alliance 162, 170
Environmental Justice Network 104
environmental sustainability 12, 64
equilibrium thinking 5, 149, 152–153, 155
equity/equality 25, 64, 69, 161
Esposito, R. 100, 104, 118
Esvelt, Kevin 57
ethico-aesthetic paradigms 4, 111, 119, 121
ethics of resilience 125, 126–127, 128, 133, 136
Evans, Brad 49
evolution 15, 181, 187
evolutionary resilience 15, 97, 149, 168
Ewald, Francois 66

Fainstein, S. 178, 187, 188
faith 5, 178, 184
faith-based communities 99, 101, 102
far right 12
Feldman, J. 24
feminism: and the body 45, 46; and community 100–101
Fidler, Ben 52
Fikes, B.J. 52
financial crisis 17, 73, 153
Fischman, Josh 50
fishing community initiative 120–121
floods/flood defences 20, 134, 140, 167, 206
Florida (US) 167
fly genetics 53–54, 56, 57, 58
Folke, C. 21, 22
Food Resilience Network 87, 89
fossil fuels 194, 195, 197, 202
Foucault, Michel 49, 115, 196, 198, 199, 200, 201
Francis, Pope 184
free market 15
Freierian pedagogy 111–112
Freud, Sigmund 186
Frith, Chris 38
Fromm, Erich 185
fruit fly (*Drosophilia Melanogaster*) 53
Fussenegger, Martin 52

Gabrys, Jennifer 33
Galileo 14
Galloway, Alexander 44, 48, 49, 55
gang violence 62
Gantz, Valentino 53

Gap Filler 87
gardening 87–88, 171
Garmezy, Norman 20
Gates, Ruth 55
GBIRd (Genetic Biocontrol of Invasive Rodents) consortium 58
Geddes, Patrick 15
gene drive technology 52–53, 56, 57, 58
genetics 52–55, 56, 58–59; and assisted evolution 54–55, 57; and fly biology 53–54; and smart bodies 45, 47–48, 52–53
Germany 22
Gibson-Graham, J.K. 83, 89
Giddens, Anthony 148
Gillis, J. 197
Glaeser, Ed 178
Glassman, James 68
Glissant, Édouard 120
global financial market 71, 72, 81, 115, 116, 117, 153, 168
global inequality 12–13, 129, 192
global society 10, 192
globalisation 104, 151, 181, 187
Goethe, Johann Wolfgang von 50
good city 5, 177–178, 182, 184
Gorz, Andre 188
governance and resilient cities 125, 126, 128, 130–136, 137–138, 140–141; citizen/stakeholder in 127, 129, 132, 138, 140; and models/typologies 131–132, *132*
governance/governmentality 7, 9, 10, 25, 111, 177; and civil society 5, 127, 135–136, 141; and communities 66, 67, 69–70; and community resilience 62; and complexity 49–50, 105, 144, 145, 147; green 203, 206; liberal 56, 67; and modernist political imaginary 110–111; neoliberal *see* neoliberalism; postmodern 155; shared 69–70; and smart bodies 44, 46, 48, 49, 56
grassroots action *see* bottom-up initiatives
Grayson, K. 96, 106
'Great Acceleration' 192, 198, 200, 204
Greely, Henry 58
green governmentality 203, 206
Greening the Rubble 87
Grenada 115
Grove, K. 83, 94, 95, 97, 100, 105
Guattari, Félix 18, 20, 47, 95, 113, 117, 119, 121
Gunderson, L.H. 21

Habermasian tradition 112
Habitat III 5, 129, 131–133
Haraway, Donna 10, 103
Harvey, D. 180, 181, 187
Hayek, Friedrich 22
Hayward, B. 85
health/healthcare 7, 45, 64, 98–99, 164
Healy, A. 154
Hegel, Georg Wilhelm Friedrich 15
Heras-Saizarbitoria, H. 73
Hernandez, Julie 23
Hill, E. 152
Hobbes, Thomas 32, 34, 200
Holling, Crawford Stanley 21, 22, 150, 194, 199–200
Hollis, L. 179
Hollnagel, E. 33
Holocene 191, 192, 196, 197, 199, 200, 202, 204, 205
homo resilis 96, 106
hope 5, 70, 83, 188
Hornborg, Alf 16
Hosking, E.N. 88
housing, affordable/public 23, 162, 163, 173, 173n5
Hudson, R. 153
human condition, The (Arendt) 181, 182–183
human rights 12, 82, 129
humanism 39, 46, 182, 183, 184, 187
Hunt, T. 182
Hunter, P. 203, 204
hurricanes 115, 116, 117, 121n1, *see also* New Orleans, disaster recovery in; Sandy, Hurricane

IBM 7
Ichinkhorloo, B. 100
ideologies 8, 17
Illich, Ivan 179, 183
images 29–30, 34, 35, 36, 38; sensory-motor/optical-sound 39–41
imagination and resilience 3, 29–41; and bioperfection imperative 39, 41; and capitalism/technology 38; and climate change 33; and critical theory 32, 33; and future disasters 31–32; and imaginary/images *see* images; and *The limits of control* (Jarmusch) 36–38; and limits of imagination 34–41; and neuroscience 37–39, 41; relations between 34; and representation 35–36, 37, 38; routinisation/bureaucratisation of 32; and selfhood 29–30, 35; and state power 32–33, 34; and subjectivities 39, 40; and survivability 30–31, 33, 34, 35; and trauma 29, 30, 31; and vulnerabilities 30, *see also* political imagination
imagined community 102–103
immigration 12, 144
individualism 15, 16, 23–24, 65, 85, 105; and smart bodies 46
industrialisation 16
inequality 16, 101, 119, 152, 153, 161, 201; bouncing back to 23; global 12–13, 129, 192
insect terrorism 58
Institute of Women and Ethnic Studies (IWES) 71
instrumentalisation of resilience 1, 5, 6, 13, 153–154; resisting 9
insurance: biopolitics of 115, 117; catastrophe *see* CCRIF; parametric 115–116, 121n1, 122n2; social 66–67, 75
interconnectivity of resilience machines 2
internet 43, 48, *see also* social media; virtual communities
Irving, D. 62
ISIS 41n1
Ivan, Hurricane 115
IWES (Institute of Women and Ethnic Studies) 71

Jacobs, Jane 180
Jacques, T. 146, 149
Jamaica 4, 100, 111, 114–117, 121n1
Jarmusch, Jim 36
Jasnoff, Sheila 8
Jay, M. 136
Jeremiah 185, 186
Jha, M. 100
Jihad, F. 146, 149
Joseph, J. 169
Joseph, M. 65
just-in-time strategy 73, 130
justice 5, 10, 13, 25, 100–101, 110, 154, 179, *see also* environmental justice

Kahn, M.E. 180
Kasarda, J. 179
Katrina, Hurricane *see* New Orleans, disaster recovery in
Katrina Memorial (New Orleans) 70
Katz, Bruce 164
Kay, L.E. 47
Kepe, T. 100

knowledge 2, 8
Kolbert, E. 195
Krafft, T. 199
Kristeva, Julia 182, 183, 186, 187
Kroker, A. 47
Kuiken, Todd 58
Kunstler, H. 181

Lakoff, A. 115
Land, C. 70, 72, 74
land, development 16, 18, 132, 172–173
land tenure 100, 140
Land Use Recovery Plan 88
land-use management/planning 17, 132, 138, 167
Langman, L. 104
language 18
Larner, W. 87, 89–90
Lash, Scott 148
Latouche, S. 204
Latour, Bruno 18, 19
Lauermann, J. 17
leadership 64, 70, 134, 154
liberal governance 56, 67
liberating potential of resilience 10–11
liminality 73, 74, 75
limits of control, The (Jarmusch) 36–38
Lindsay, G. 179
liquid modernity 6
local government 70, 84–85, 88–89, 90
locale *see* place-based communities
Locke, John 15, 200–201
Logan, John 16, 24
longevity 33, 45, 59
Louisiana Justice Institute 24
Luhmann, Niklas 22
Luke, T.W. 199, 200
Lyttelton, Project 88

McCardle, A. 161
machines/machine metaphor 1, 3, 6, 14–18, 80; diagrammatic/autopoietic 113; and evolution 15; and mechanistic universe 14–15; and technology 16; and urban politics 16–17
machinic assemblages 20, 24, 25, 81, 89, 106, 113, 117
MacKinnon, D. 153
MacQueen, K.M. 98–99
'making room for water' strategy 5, 160, 166–167
malaria 54, 56, 57, 58
Malins, P. 18

managerialism 3–4, 70, 72
Manhattan (New York) 160, 161, 163, 165, 174n7
Manson, S. 19
market managerialism 63, 72, 73–74, 75
market relations 69, 75
markets/market economy 15, 85, 105, 125, 128, 141, 203; carbon emissions 169, 202; and community 63; and corporations 13, 17, 206; disequilibria 205; global/financial 71, 72, 81, 115, 116, 117, 153, 168; and governance 137–138; labour 162
Martin, R. 151
Marx, Karl 185, 186, 201
Marxism 111–112
Marzec, R.P. 198, 199
mass extinction 9–10
Massey, Doreen 101–102
Massumi, Brian 8, 18, 56
MCC (Mondragon Cooperative Corporation) 72–73, 74
mechanistic universe 14–15
Meira, F. 73–74, 75
Melbourne (Australia) 177
Microsoft 7
Millennium Development Goals 129
Miller, P. 138
Minton, L. 43
MIT (Massachusetts Institute of Technology) 53, 55, 57
modernist political imaginary 110–111, 112
modernity 65, 179, 180–181, 183, 185; deformed 200–202
Mohammed, Khalid Sheikh 32, 38
molecular manipulation *see* smart bodies
Molotch, Harvey 1, 16–17, 18, 24
Morton, Adam 35, 37
mosquitoes 53–54, 56, 57, 58
Moss, T. 22
Mummery, J. 104
Murphie, A. 121
Myers, N. 50

Nancy, Jean Luc 65, 104
natality 177, 178, 182, 183
natural resources 100, 104–105, 188
Nelson, S.H. 4, 83, 86, 88
Neocleous, Mark 32–33, 41n3, 57
neoliberalism 1, 2, 13, 126, 153–154; and the Anthropocene 191–192, 194, 195–196, 198, 206; and autonomisation 63; and bottom-up initiatives 8; and

communitas 73–74; and communities *see under* community resilience; and imagination 38; and instrumentalisation of resilience 1, 5; and managerialism 4, 63, 70–71, 72, 73–74, 75; and post-disaster recovery 63, 80, 81–82; and resilient cities 127, 128, 130, 137–138, 139, 140–141; and resilient subjects 67; and smart bodies 45, 47, 52, 57; and social welfare 68; and spontaneous order 22; and taxation 169; and TINA 3, 63
nescience 148
Neslen, A. 58
networks 48–50, 64, 66, 149, 171; as anarchic/anti-authority 48; and freedom/flow 48–49; and political subjection 49; and protocol/governmentality/security 48, 49; smart bodies as 45–46, 48
neuroscience 37–39, 41
New Jerusalem 182
New Orleans, disaster recovery in 3–4, 23, 63, 68–71, 72, 74, 171; and equity/social capital 69; and exclusion 68, 69, 71; and extent of disaster 68; and IWES 71; as opportunity for transformation 68, 71; resistance narrative in 68–69; and resistance to resilience 24–25; and UNOP/'green dot' plan 69–70, 71; and welfare dependency 68
'New Urban Agenda' programme 129, 130, 131–133
New York City Environmental Justice Alliance (NYC-EJA) 162, 170, 171–172, 173
New York City, resilience planning in 159–173; and adaptation 166–170; and affordable housing 162, 163, 173, 173n5; and Bloomberg/*PlaNYC* 159, 160–161, 164, 170, 171; and community organisations 170–172; de Blasio plan/*One/NYC* 161–163, 164–165, 169–170, 173n5; and environmental justice 5, 160, 161–162, 164–166, 171–172, 173; and Hurricane Sandy (2012) 5, 159, 170; 'making room for water' strategy 5, 160, 166–167; and neoliberalism 166, 168, 169; and parks 169–170, 172; Progress Report (2016) 163; and rising seal levels 159, 162, 163, 172; and sustainability 160, 161, 162, 165, 170; and taxation 169; and waterfront land 160–161, 162, 163, 166–167, 169–170

New York Regional Assembly 171
New York Times 12, 52–53, 68
New Zealand *see* Christchurch, disaster recovery in
Newton, Isaac 14
NGOs (non-governmental organisations) 7, 134, 139
Nightingale, A.J. 168
9/11 attacks 31, 32, 163
North Carolina State University 58
NYC-EJA *see* New York City Environmental Justice Alliance
Nytray, Crystal 51

Olsson, P. 147
'100 Resilient Cities' 5, 7, 13, 13–15, 85, 129, 130, 133–135; and CROs 134–135; in New York 159–160
OneNYC 161–163, 164–165, 170
Ong, A. 102
ontology of potentiality 4, 83, 86
organic world 14
organisation theory 147
organisational cultures 128, 139
organisational studies (OS) 3, 63

Palomino-Schalscha, M. 88
panarchy model 21–22, 154, 155, 156n1, 206
pandemics 57
Pareto, Vilfredo 22
Parker, M. 63, 74
Parnet, C. 19
Parsons, Talcott 22
participatory adaptation/resilience 110, 111, 112, 113, 119, 120–121
participatory planning 69–70, 112
Pearl Harbor attack (1941) 31–32
Peck, J. 89
Pelling, M. 82
physics 15, 16
place-based communities 99, 101–102
PlaNYC 159, 160–161, 164
Plato 34
Point Community Development Corporation (CDC) 171
political imagination 32–34, 41n3
political machines 16
politics of resilience 22, 29, 107, 110–111; modernist imaginary of 110–111, 112; necro- 59; and organising 63; and procedural/distributional struggles 112
pollution 169, 171, 173n4, 174n7, 180

Porter, L. 22
possibility spaces 8, 24, 25
post-capitalism 81–84, 183; and post-politics 82, 86, 87, 89–90; and transformation 83–84
post-disaster recovery 3–4, 9, 62, 64, 140; and bounce-back-ability see bounce-back-ability; and community-led/transformational projects 84, 87–89, 90; and depoliticization 82, 83, 86, 89; and encroachments on democracy 82, 85, 89, 90n1; and neoliberalism 63, 80, 81–82, 85; and post-capitalism see post-capitalism; and post-politics 82, 86, 87, 89–90; and regressive policies 82; and self-responsiveness 82, 85–86; and transformation 68, 71, 81, 83–84, see also Christchurch; New Orleans
post-modernity/-modernism 105, 145, 145–146, 148, 155
post-politics 82, 86, 87, 89–90
poverty 40, 56, 62, 71, 129, 162, 180
Povinelli, Elisabeth 9
power 2, 8, 10, 37, 112, 204; and community 105–106; resistance to see resisting resilience machines; and self 30; and smart bodies 48, see also empowerment
power relations 18, 19, 82, 100, 151–152, 194
pre-emptive thinking 56–57
Prince, Samuel 84
property rights 15
protocol 48, 49
protological control 44, 46, 47–50
psychology 20, 23, 29, 31, 33, 41, 64, 196
public good, resilience as 17, 22
public health see health/healthcare
public housing 23, 161, 162
Putnam, Robert 65–66

Radkau, J. 200
Rancière, Jacques 95
RAND Corporation 62, 69, 71
Randalls, S. 96
Rankine, William 20
reason 5, 14, 46, 184, 185, 186, 187
refugees 13, 144
Regaldo, A. 52, 55, 57, 58
regional planning 7, 145
Reid, Julian 49
Reid, R. 96
religious communities 99, 101, 102

religious thinking 182–184
representation 35–36, 37, 38
resilience 1–11; aesthetics of 95, 97, 98, 118; ambiguities/fluidity of 7, 8, 13–14, 97, 125, 149; and critical political geography see critical political geography; critiques of 3–4, 6, 9, 75–76, 94–98, 125, 126; ecological sense of 21; ethics of 125, 126–127, 128, 133, 136; evolutionary 15, 97, 149, 168; framed as apolitical 82, 154; genealogy of 20–25, 97; as governmentality 25; interdisciplinary nature of 149, 155, 168; and justice see justice; and knowledge/control 2, 8; multiple nature of 96–97; and neoliberalism see neoliberalism; origin of term see resilience, genealogy of; as polysemic concept 129, 130; preponderance of 1, 6, 9, 13, 21, 94, 196–197; psychologists of 29; remoteness from daily life of 7; resisting see resisting resilience machines; and sustainability 12; and transformation see transformation; without hope 5, 178, 187–189
Resilience Alliance 13, 21, 129, 154, 192
resilience capacities 138, 139, 140
resilience coalitions 1, 5, 151
resilience diagram 95–98, 106; and aesthetics of resilience 95, 97, 98; bounded/unbounded form 98, *98*, 107; closed/open spectrum form 97, *97*, 107; and multiple nature of resilience 96–97
resilience, genealogy of 3, 20–25; in ecology 21, 22, 23; in engineering 21–22; and individualism/self-reliance 23–24; as normative concept 22–23; in psychology 21, 23; in social science 22; and systems/complexity theory 21–22
resilience implementations 4–5
resilience planning see New York City, resilience planning in
resilience policy 4, 9, 13, 192; global frameworks 125, 129–136; and practice 125, 126
resilience research/studies 63, 64, 71, 76, 119; paradox in 110–111
resilience science 192, 194, 196, 197–198, 199–200, 203, 204, 205
resilience strategies 4–5, 7, 125, 126, 127; building 137–138; and global policy frameworks 129–136; hurdles/dangers in 138; and neoliberal governance

129–130; and scale/scope 125–127, 128, 130, 131, 133, 137, 141
resilience thinking 10, 13, 20, 50; and bioengineering 56–57; and catastrophe 195, 197; and community 94–98, 105, 106–107; conceptual shortcomings of 22; and diagrams/maps *see* resilience diagram; paradoxes of 21–22; and strategy/practice 127, 130, 133, 135, 136, 137–138, 140
resilience-building technologies 3
resilient cities 17, 125–141, 151, 177; and '100 Resilient Cities' *see* '100 Resilient Cities'; citizen/stakeholder in 127, 129, 132, 138, 140; and civil society 5, 127, 135–136, 141; and disaster management 127, 130, 131, 140; global policy frameworks for 125, 129–136, 139; and governance *see* governance of resilient cities; and Habitat III/'New Urban Agenda' 129, 130, 131–133; and land-use management/planning 132, 138; and models/typologies 131–132, *132*; and neoliberalism 127, 128, 130, 137–138, 139, 140–141; and policy/strategy/practice 125, 126–127; and resilience capacities 137, 138, 139, 140; and resilience as process/governing technique 126–127, 128, 137; and resilience thinking 127, 130, 133, 135, 136, 137–138, 140; and scale/scope 125–127, 128, 130, 131, 133, 137, 141; and SDGs 129, 130, 131, 133; strategies for 125, 126, 127–128; and sustainable development 127, 129, 130, 131, 136; and third sector organisations 129, 133, 138; UN Report on (2004) 24
resilient citizen/citizenship 85, 127
Resilient Melbourne 177
Resilient New Orleans 69
resilient subjectivities 1, 2–3, 6, 8, 9, 13
resilient subjects 49, 67, 83, 126, 127, 200–202, 206
Resistance Alliance 13
resisting resilience machines 8–9, 10, 18; and alternative work organisations 71–74, 75–76; and community-led/transformational projects 84, 87–89, 90; and danger of co-option 8, 82, 89, 165; and everyday/local scale 80, 81; 'Stop calling me resilient' campaign 24–25
restoration engine 6, 192, 193, 196–206; and catastrophism 192–194; and hidden recoil of resilience 193–194, 202, 203, 204, 205–206; and modernity 200–202; shortcomings of 202–204; in systems of systems 196–200
risk 2, 16, 23, 82, 85, 112; assemblages 115; and planning 167–168; and social insurance 66–67, *see also* disaster risk management; vulnerability
risk management 13, 24, 66, 113, 115, 116, 130, 206
Rittel, Hans 146
Roberts, N. 200
Rockefeller Foundation 161, 164, *see also* '100 Resilient Cities'
Rodin, Judith 160
Rose, Nikolas 56, 66, 67, 74–75, 138
Rousseau, J.-J. 201

Salt, D. 2, 83, 192, 197, 198, 199
Sandhu, K. 100–101, 102
Sandy, Hurricane (2012) 5, 159, 170
Sarewitz, D. 204
scale 3, 18, 80, 151; everyday/local 80, 81, 94; interdependencies of 2, 25; molecular *see* molecular manipulation; and resilient cities 125–127, 128, 130, 131, 133, 137, 141
Schmid, C. 198
Scientific Revolution 14–15
SDGs (Sustainable Development Goals) 129, 130, 131, 133
Second World War 31–32, 182
security 32–33, 41n1, 43
self, resilient *see* imagination and resilience
self-determination 129, 138, 140
self-help 24, 30
self-reliance/-responsiveness 23–24, 82, 85–86
self-responsibility 73, 88
self-sufficiency 24, 64, 68
Shakan, Jeremy 197
shared governance 69–70
Shimoda earthquake (1854) 20
significations 18
Sikhs 101
Simmie, J. 151
Simon, S. 96
Slaby, Jan 39
slums 180
smart bodies 3, 44–59; and anti-microbial resistance 57; and auto-poetic nature 44, 46, 47, 51, 55, 56; and biological sovereignty 46, 48, 49, 50; and biomimicry 44–45; and cellular manipulation 51–52; and code

perspective 47; and coral reefs/assisted evolution 54–55, 57; and corporations 59; and CRISPR/gene drive technologies 52–53, 56, 57, 58; dangers of 45, 46, 57–58; and diabetes treatments 51; and eugenics 57, 58; and feminist scholarship on body 45, 46; and fly genetics 53–54; and governance/power 44, 46, 48, 49, 56; and human–non-human separation 47; implications of 56–59; and individuality/agency 46, 47; and military 58; and neoliberalism 45, 47, 52, 57; as networked life 45–46, 48; as paradoxes of 44; as political/preventative strategy 55, 56–57; and protological control 44, 46, 47–50; as self-organising 44, 49, 51; and technomimicry 45; theorisation of 46–47; and tissue engineering 50–51
smart drugs 52
smart technologies 3, 7, 43, 45; and interoperability 43; and self-organisation 44, *see also* smart bodies
Smith, M.P. 17
social capital 64, 65, 69, 102
social contract 111, 201
social insurance/risk 66–67
social media 41*n*1, 69, 103
social networks 64, 66
social relations 18, 62–63, 81, 103, 104
social resilience 95, 104–105, 187
social science 13, 15, 18, 22; 'non-human turn' in 45; resilience in 20, 21, 22, 33, 39
'social, the' 22, 66, 74–75
social-ecological systems 150, 166, 168, 197, 198
sociology 22, 65, 66
Soguk, Nevzat 10
solidarity 63, 65, 67, 74, 75; degeneration of principles of 72–73; and social insurance 66
Sonderhaus, F. 22
South Africa 100
sovereignty/sovereign subject 110, 111, 113, 118, 119; biological 46, 48, 49, 50
Spadaro, A. 184
Spain 72–73
spontaneous order 22
stability 6
state power 3; and imagination 32–33, 34
state, social 66, 67
Stockholm Resilience Centre 129
Storey, J. 73

strategic framing 70, 126, 128, 130, 137, 140, 141
Streeck, W. 181, 188
Strega, S. 9
subjectivities 1, 2, 6, 111, 113, 119; and imagination 39, 40; and smart bodies 47
sustainability/sustainable development 12, 64, 127, 129, 130, 131, 136, 197
Sustainable Development Goals (SDGs) 129, 130, 131, 133
Swanstrom, T. 23
Swiss Federal Institute of Technology (ETH) 52
Swyndegouw, Eric 168
Synlogic 52
systems theory 2, 21–22, 99, 147, 149, 152, *see also* complex systems

Tampio, N. 19
taxation 169
technology 16, 38, 195
technomimicry 45
terrorism 12, 58, 193; and radicalisation 62
Thacker, Eugene 44, 48, 49, 55
thermodynamics 15, 16
third sector organisations 129, 133, 138
threat *see* vulnerability
thresholds 9, 198
Thrift, N. 20
Tickell, A. 89
Tidball, K. 88
timebanks 88, 90*n*2
Timmerman, L. 52
TINA (There Is No Alternative) 3, 63
tissue engineering 50–51
Tobin, G.A. 23
Total Quality Management 73, 74
transformation 68, 71, 126–127; and post-disaster recovery 81, 83–84, 87–89, 90
Transition Network 103
Transitional Architecture movement 87
translocal communities 102
trauma 29, 30, 31, 35, 88
Turner, Victor 73

uncertainty 5, 9, 21, 64, 81, 82, 184; and urban development 144–147
United Nations (UN) 24, 178; Office for Disaster Risk Reduction 13; SDGs 129, 130, 131, 133
United States (US) 16, 17, 18, 205, 206; and 9/11 attacks/War on Terror 31–33,

38; community resilience programmes in 62, 66, 75, *see also* New Orleans, disaster recovery in; New York City, resilience planning in
UNOP (Unified New Orleans Plan) 69–70
urban agriculture 87–89
urban development 70, 141; as non-controllable 145–147
urban growth machine 1, 16–18, 125, 169; and resilience machine, compared 17–18, 193
urban imaginaries 5, 178
urban planning 5, 21, 144–155; and adaptive cycle model 150, 151, 154, 155; and complexity/uncertainty 144–147, 152, 154; and critical political geography *see* critical political geography; and equilibrium thinking 149, 152–153, 155; and nescience 148; and post-modernity 145–146; rational 146, 156n3; and resilience machines 145, 151–152, 155; and system thinking/resilience 148–150; as wicked problem 145, 146, 147, 155, *see also* New York City, resiliency planning in
urban politics 16–17, 18; and experimentation 17
urban resilience 5, 9, 95, 125, 144, 145, 155; ambiguities in research on 149, 155; capacities approach to 138, 139; and digital technologies/big data/algorithms 7; in global policy 129–136, 140
urban resilience machine 127, 128, *see also* resilient cities
urbanisation 2, 125, 178–180, 182, 204; and accumulation 179; and alienation/separation from nature 185–186; and 'good enough city' 178, 184–187; and humanism 182, 183, 184, 187; and slums 180
utopian thinking 182, 184, 188, 201

van Munster, R. 84
Venter, C. 56
Victorians 182
virtual communities 97, 103–104, 107
vital systems security 115
vulnerabilities 2, 3, 11, 22, 82, 85, 100, 105, 111; depoliticisation of 110, 112–113; of human bodies 6; of self 30

Wakefield, Stephanie 10
Walker, B. 2, 83, 192, 197, 198, 199
War on Terror 33, 56
Washington, Tracie 24
water insecurity 22, 58, 160
waterfront land 160–161, 162, 163, 166–167, 169–170
Weber, Max 65
welfare state/systems 18, 23, 67, 68, 75
Welsh, M. 144, 152, 153
White, I. 167
wicked problems 125, 145, 146, 147, 155
Wilkinson, C. 152
Williams, R. 65, 99
worker cooperatives 72–73
World Bank 13, 115
World Economic Forum 12
World Social Forum 104

Yee, Aubrey 57, 58, 59
Yeh, E.T. 100
Young Rohjan, S. 52
youth development 64
Yusoff, Kathryn 33

Zebrowski, Chris 49, 56
Zhang, G. 51
Zika virus 54
Zimmer, C. 53, 57
Žižek, S. 178, 179, 180, 198
Zuboff, Shoshana 38

CPSIA information can be obtained
at www.ICGtesting.com
Printed in the USA
LVHW011705080119
603165LV00019B/869/P